More MATHEMATICAL Quickies & Trickies

Over 300 Non-Routine Questions
to Enhance Your Problem-Solving Skills

Yan Kow Cheong

MATHPLUS
PUBLISHING
Specialists in Mathematics Education

For P6 and above

MATHPLUS Publishing
Blk 639 Woodlands Ring Road
#02-35 Singapore 730639

E-mail: publisher@mathpluspublishing.com
Website: www.mathpluspublishing.com

First published in Singapore in 2011

Copyright © 2011 by Yan Kow Cheong

All rights reserved. No part of this publication may not be reproduced, stored in a retrieval system, or transmitted in any form or by any means, electronic, mechanical, photocopying, recording or otherwise, without the prior written permission from the publisher.

National Library Board, Singapore Cataloguing-in-Publication Data

Yan, Kow Cheong.
    More mathematical quickies & trickies / Yan Kow Cheong. – Singapore : MATHPLUS Pub., 2011.
    p. cm.
    ISBN : 978-981-08-5413-3

    1. Mathematical recreations. I. Title.

QA95
793.74 -- dc22   OCN671581211

Printed in Singapore

PREFACE

More Mathematical Quickies & Trickies, the long-awaited sequel to **Mathematical Quickies & Trickies** (first published in 1998, and recently expanded and updated) similarly adopts a fun and humorous approach to mathematical problem solving.

For those of you who are not familiar with the terms **mathematical quickie** and **mathematical trickie**, let me define them again:

A **mathematical quickie** is a problem which may be solved by laborious methods, but which with proper insight may be disposed of quickly. This term was coined by the late Professor Charles W. Trigg to describe problems that yield almost instantly to a flash of inspiration.

A **mathematical trickie** is a problem whose solution rests on some key word, phrase, or idea rather than on a mathematical routine. Most number riddles would thus qualify as trickies.

Over 300 elementary quickies and trickies are compiled, from the fields of arithmetic, number theory (higher arithmetic), geometry, algebra, and recreational mathematics, to challenge both upper primary and secondary school (grade 5-10) students. Most problems should prove accessible to an above-average primary school student, but a number of them might challenge even a talented or gifted secondary student.

Besides the 300 odd trick and tricky questions, **More Mathematical Quickies & Trickies** has also more than 25 five-minute enrichment mathematics mini-articles to enliven the spirit of this lightly irreverent approach to mathematical problem solving.

It is hoped that **More Mathematical Quickies & Trickies** will not only expose readers to different problem-solving techniques, commonly used in answering math contests and competitions questions, but will also encourage them to come up with their own elegant or intuitive solutions other than the solutions provided in this sequel.

Once again, I look forward to hearing from readers who would like to share with me their ingenious solutions. You may contact me at kcyan.mathplus@gmail.com and visit my blog at http://singaporemathplus.blogspot.com or follow me on Twitter as @MathPlus.

K C Yan

CONTENTS

	Preface	iii
1	Creative GST	1
2	Are You Calculator-Smart?	9
3	What Is the Easy Way?	17
4	The Magic of 3 Consecutive Numbers	23
5	Twitter Math @MathPlus	28
6	What Is 27 × 37, Really?	35
7	Humanizing 1, 2, 3	42
8	A Mathophobia Kit	48
9	WITs – 13 Ways to Attain Mathematical Excellence	55
10	Facebook Math: Numeracy vs. Literacy	61
11	*Thou Shalt Not Divide By Zero*	68
12	Math Jokes to Relieve Stress	75
13	*Look-see* Proofs	82
14	Some PhD Math Questions	90
15	Mathematical Prayers	95
16	The Largest Product	102
17	What's Wrong? – A Comedy of Mathematical Errors	109
18	The Aha! Myth	117
19	Sam Loyd's *Toughies*	124
20	The Tuesday Boy Problem	136
21	What Is 1 + 1, Really?	143
22	In Love with Cryptarithms	151
23	*Mathematical Kiasuism*	156
24	The Mathemagic of 142857	161
25	The Lighter Side of Singapore Math	167
26	K C Yan's Laws & Lores	167
27	Flee and Free from the FREE	173
	Answers	181
	Bibliography & References	204

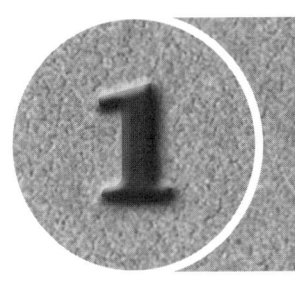

1 More Mathematical Quickies & Trickies

Creative GST

How do you effectively work out 7% GST (Goods & Services Tax)?

Work out 7% GST on a bill of $15.

Let's see how four students mentally arrive at the answer. Which method do you think is better?

Jill said:

20% is $\frac{1}{5}$ and $\frac{1}{5}$ of $15 = $3
Dividing by 10 gives 2% which is 30¢
10% of $15 is $1.50
5% is $\frac{1}{2}$ of $1.50 which is 75¢
7% of $15 is 30¢ + 75¢ = $1.05

Ryan said:

10% of $15 is $1.50
1% is $\frac{1}{10}$ of $1.50 = 15¢
3% is 3 × 15¢ = 45¢
7% is 10% − 3% = $1.50 − 45¢ = $1.05

Mental calculation provides a rich opportunity to be creative!

Henry said:

7% of $15 is the same as 15% of $7.
5% of $7 is 35¢ so 15% of $7 is 70¢ + 35¢ = $1.05

Marie said:

50% of $15 is $7.50
1% of $15 is 15¢
49% is $7.50   15¢ − $7.35
7% is $7.35 ÷ 7 = $1.05

The calculations show four different mental strategies used in working out 7% GST on a bill of $15. Of course, there is a range of other strategies which may be used in finding the answer. In other words, a simple calculation such as finding the GST provides a rich opportunity to creative mental calculation.

Practice

1. You are visiting the United Kingdom, and you need to mentally work out $17\frac{1}{2}\%$ VAT (Value Added Tax) on an item costing $48. Give three different ways of doing the calculation.

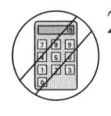

2. While visiting a tourist attraction site, you plan to buy a souvenir item costing $30. The sales tax is 8%. Think of four ways of doing the calculation.

Answers

1. **Method 1**

    $17\frac{1}{2}$ of $48 = 35\%$ of $24
    $= 70\%$ of $12
    $= $8.40

    Method 2

    10% of $48 = $4.80

    $\frac{1}{2}$ of it is 5% = $2.40

    $\frac{1}{2}$ of it again is 2.5%

    So $2\frac{1}{2}\%$ is $1.20

    Altogether that is $4.80 + $2.40 + $1.20 = $8.40

    Method 3

    $12\frac{1}{2}\%$ of $48 is $\frac{1}{8}$ of $48, which is $6

    5% is $2.40

    Altogether, it is $8.40.

2. **Method 1**

    20% is $\frac{1}{5}$, and $\frac{1}{5}$ of $30 is $6.

    Divide by 10 to give 2%, which yields 60¢
    Double that twice gives 8% which is $2.40

    Method 2

    10% is $3 so 2% is 60¢.
    Now, 8% is 10% − 2% which gives $3 − 60¢ = $2.40

    Method 3

    8% of $30 is the same as 4% of $60, or 2% of $120, or 1% of $240 and that is $2.40.

    Method 4

    8% of $30 is the same as 30% of $8.
    10% of $8 is 80¢ so 30% of $8 is $2.40.

1 When Mr. Yan collected the newsletter page 13 was missing. If the back page was numbered 20, what other pages were missing?

2 Around a fixed wheel, another wheel is rolled such that it makes four full turns in one complete revolution. Find the ratio of the radius of the rolling wheel to that of the fixed wheel.

> Mr. Yan: What is 9 × 6?
> Student: 54!
> Mr. Yan: Great! And what is 6 × 9?
> Student: 45!

3 If five frogs can catch five flies in five minutes, how many frogs are required to catch a hundred flies in a hundred minutes?

4 At what time are the hands of the clock pointing in exactly opposite directions, each towards a minute division line?

5 What is the last digit of 7^{7777}?

$7^{7777} = \underbrace{7 \times 7 \times \ldots \times 7}_{7777 \text{ times}}$

---

6 Find the next term in the series:

$$4, \quad 10, \quad 25, \quad \underline{}.$$

Note that several valid answers are possible. What would be the expected one?

7 A fence enclosing the greatest possible rectangular area is to be built. Find the ratio of the length to its width if 10 meters of fencing is available.

8 A total of 28 handshakes were exchanged at a Mathematics meeting. Each person shook hands exactly once with the other members. How many members members were present at the meeting?

> Mrs. Lou: What is $2k + k$?
> Student: 3000!

9 Find all solutions of the equations: $xy = yz = zw = wx = 1$.

10 It takes $1\frac{1}{2}$ men $1\frac{1}{2}$ days to plant $1\frac{1}{2}$ mathium in $1\frac{1}{2}$ pots. How much will $2\frac{1}{2}$ men plant in 9 days?

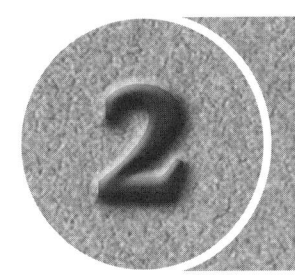

2 More Mathematical Quickies & Trickies

Are You Calculator-Smart?

Using your calculator, how would you work out the following?

(a) **Two consecutive numbers whose product is 2256.**

(b) Three consecutive numbers whose product is 12,144.

(c) **Two consecutive numbers whose squares have a sum of 2813.**

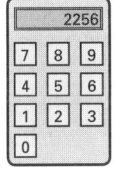

(d) Two 2-digit numbers whose product is 1219.

(e) **Two 2-digit numbers whose quotient is 0.453 125.**

(f) Two numbers with a sum of 100 and a product of 1771.

(g) **Three numbers with a sum of 100 and a product of 35,964.**

Solution

(a) By trial and error we could find that $40 \times 41 = 1640$ and $50 \times 51 = 2550$. So we then try some pairs in between. And to reduce the search we look for pairs which will give 6 as units (or ones) digit.

A better strategy is to find the square root, which lies between the two consecutive numbers.

$$\sqrt{2256} \approx 47.497 \text{ and so } 47 \times 48 = 2256$$

Hence the consecutive numbers whose product is 2256 are 47 and 48.

(b) Let's find the cube root of 12,144, which should be close to any of the three numbers.

$$\sqrt[3]{12,144} \approx 22.985 \text{ and so } 22 \times 23 \times 24 = 12,144$$

Hence the three consecutive numbers whose product is 12,144 are 22, 23, 24.

(c) To find two consecutive squares whose sum is 2813, let's find half of 2813 and then look for the squares immediately above and below.

$$\frac{1}{2} \times 2913 = 1406.5$$

$$\sqrt{1406.5} \approx 37.503$$

Now, $37^2 + 38^2 = 2813$

So the two consecutive squares are 37^2 and 38^2.

(d) Expressing 1219 as a product of prime factors, we have

$$1219 = 23 \times 53$$

Hence, the two 2-digit numbers whose product is 1219 are 23 and 53.

> Note that since $\sqrt{1219} \approx 34.9$, one of the factors must be below 34.

(e) Two 2-digit numbers whose quotient is 0.453 125.

$$0.453\ 125 = \frac{453{,}125}{1{,}000{,}000}$$

$$= \frac{453{,}125 \div 125}{1{,}000{,}000 \div 125}$$

$$= \frac{3625}{8000}$$

$$= \frac{3625 \div 125}{8000 \div 125}$$

$$= \frac{29}{64}$$

Hence the two 2-digit numbers are 29 and 64.

(f) By prime factorization,
$$1771 = 7 \times 11 \times 23 \text{ and } 77 + 23 = 100$$

So the two numbers are 23 and 77.

(g) $35{,}964 = 2 \times 2 \times 3 \times 3 \times 3 \times 3 \times 3 \times 37$
$\phantom{35{,}964\ } = \underbrace{2 \times 2 \times 3 \times 3}_{36} \times \underbrace{3 \times 3 \times 3}_{27} \times 37$
$\phantom{35{,}964\ } = 36 \times 27 \times 37$

$36 + 27 + 37 = 100$

Hence the three numbers with a sum of 100 and a product of 35,964 are 27, 36, and 37.

> Find two 2-digit numbers whose quotient is 0.589 744 rounded to 6 decimal places.

1 A cassette has 90 minutes of playing time. If you need to record four compact discs with playing times of 92, 88, 95, and 101 minutes, how many cassettes will be needed, and how much blank tape will be unused?

2 The sequence 60, 60, 24, 7, 52 is a series of unrelated numbers, yet the numbers are connected in some way. How are they related then?

3 Abacusi and Sorobahy are 200 km apart, linked by a straight road. Abram lives exactly halfway along that road. He has to meet his lawyer in Abacusi at 12:12 p.m. the next day. If Abram leaves home at 10:00 a.m. and drives at a steady 50 km/h, will he arrive in time?

4 The diagram shows a circular table in the corner of a room, touching a 4 feet by 6 feet rectangular piece of furniture. Find the diameter of the table.

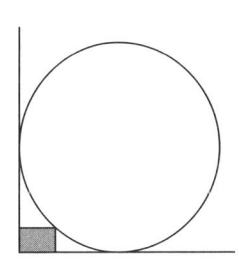

5 A tank contains exactly one liter of a 90% acid-water mixture. How much water must be added to change the liter to an 80% acid mixture?

6 (a) Draw only four straight lines such that they pass through all nine dots.
  *(b) Draw only three straight lines such that they pass through all nine dots.

○ ○ ○
○ ○ ○
○ ○ ○

7 Compute the value of the following equation:

$$\frac{1234567}{1234568^2 - (1234567 \times 1234569)}$$

8 One invention saves 30% on fuel; a second, 45%; and a third, 25%. If you use all three inventions at once can you save 100%. If not, how much?

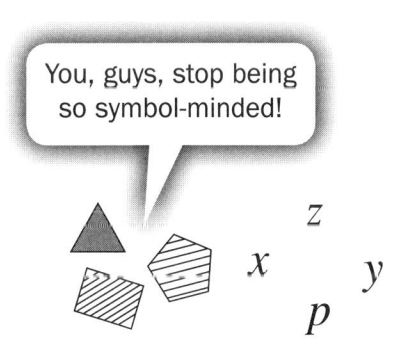

You, guys, stop being so symbol-minded!

x z
p y

9 Robin's father has three children. One is Betty, who lives in Singapore. Another is Arthur, who lives in Malaysia. Who is the third?

10 A plywood sheet is 45 inches by 45 inches. What is the approximate diameter of the log the sheet was made from? ($C = \pi D$)

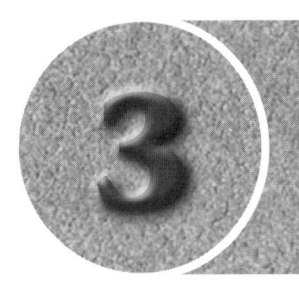

3 More Mathematical Quickies & Trickies

What Is the Easy Way?

Without a calculator, find an easy way to do the following.

(a) To add 19 to a number.

(b) To subtract 19 from a number.

(c) To add 99 to a number.

(d) To subtract 99 from a number.

(e) To multiply a number by 8.

(f) To divide a number by 8.

(g) To multiply a number of 25.

(h) To divide a number by 25.

(i) To multiply a number by 99.

(j) To divide a number by 99.

Always look out for shortcuts when operating on numbers.

Solution

(a) To add 19 to a number, add 20 and subtract 1.

For example, $47 + 19 = 47 + 20 - 1$
$= 67 - 1$
$= 66$

(b) To subtract 19 to a number, subtract 20 and add 1.

For example, $82 - 19 = 82 - 20 + 1$
$= 62 + 1$
$= 63$

(c) To add 99 to a number, add 100 and subtract 1.
$$176 + 99 = 176 + 100 - 1$$
$$= 276 - 1$$
$$= 275$$

(d) To subtract 99 from a number, subtract 100 and add 1.
$$345 - 99 = 345 - 100 + 1$$
$$= 245 + 1$$
$$= 246$$

(e) To multiply a number by 8, double, double, and double again.
$$37 \times 8 = \underbrace{37 \times 2} \times 2 \times 2$$
$$= \underbrace{74 \times 2} \times 2$$
$$= 148 \times 2$$
$$= 296$$

(f) To divide a number by 8, halve, halve, and halve again.
$$256 \div 8 \rightarrow \frac{256}{2} = 128 \rightarrow \frac{128}{2} = 64 \rightarrow \frac{64}{2} = 32$$

So $256 \div 8 = 32$

(g) To multiply a number of 25, multiply by 100 and divide by 4.
$$46 \times 25 = \frac{46 \times 100}{4} = \frac{4600}{4} = 1150$$

(h) To divide a number by 25, divide by 100 and multiply by 4.
$$1700 \div 25 = \frac{1700}{100} \times 4 = 17 \times 4 = 68$$

(i) To multiply a number by 99, multiply by 100 and subtract the number.
$$56 \times 99 = 56 \times (100 - 1)$$
$$= 5600 - 56$$
$$= 5544$$

(j) To divide a number by 99, divide by 9 and then by 11.
$$2376 \div 99 \rightarrow 2376 \div 11 = 216 \rightarrow 216 \div 9 = 24$$
$$2376 \div 99 = 24$$

1 Calculate 999,999,999 × 12,345,679.

How I can solve this without a calculator?

---

2 A clock reads 2:00 on Sunday afternoon. If it is running slow, and losing 9 minutes every six hours, what time will the clock show when it is actually 8:00 on Monday evening?

3 If each letter of the alphabet is worth three times its order value, with A as number 1 and Z as number 26, what is the total value of all the letters in the alphabet?

4 If STOP equals 70, what does START equal?

　　A. 56　　　B. 74　　　C. 78　　　D. 80　　　E. 83

5 An ant starts walking from one corner of a cube by confining herself to its edges, which are all exactly 1 cm long. If the ant walks as far as she could possibly can without retracting any part of her path, how far does she walk?

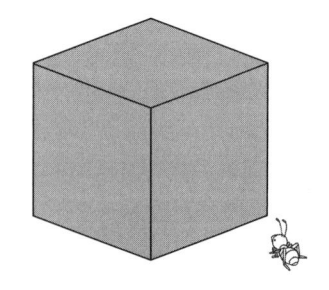

6 What is the remainder when 10^{99} is divided by 9?

7 An *idiot savant* goes up the stairs 5 steps and comes down 4 steps. He counts one for each step he goes up, but ignores to count any steps he goes down. By the time he reaches the top of the floor, he has counted 45 steps. How many steps are there?

8 The digits 0, 1, 2, 5, 6, 8 and 9 on a calculator appear unchanged when it is turned upside down. For example, the number 62,528 remains the same when the display is upside down. How many whole numbers between 100 and 1000 remain the same when the calculator is turned upside down?

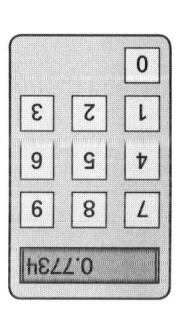

9 A pencil costs 22 cents, a pen 28 cents, and an exercise book 37 cents. What is the least whole number of dollars one needs to spend to get at least one of each item without any change?

10 At a meeting of 24 members, 17 like algebra, 13 like geometry, and 8 like calculus. What is the smallest number of members who like more than one of the subjects?

Algebra |||
Geometry ||||
Calculus |||| ||

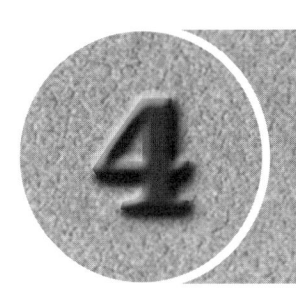

4 More Mathematical Quickies & Trickies

The Magic of 3 Consecutive Numbers

Consider three consecutive numbers, say, 6, 7, 8. Compare the square of the middle number with the product of the outer pair.

$$6 \times 8 = 48$$
$$7 \times 7 = 49$$

The answers, 48 and 49, are also consecutive.

What about the numbers 12, 13, 14?

$$12 \times 14 = 168$$
$$13 \times 13 = 169$$

It works again.

Verify that it works for the triples (4, 5, 6) and (11, 12, 13).

Is the pattern always true for any consecutive three numbers?

Algebraically speaking, if we denote the middle number by n, then the difference of two squares is

$$n^2 - 1 = (n-1)(n+1)$$

which explains why the pattern holds true.

Extension
1. Use a similar argument to show that three consecutive odd numbers or three consecutive even numbers will yield answers that differ by 4.

2. Show that multiples of 3 or other arithmetic sequences where the common difference is 3 will give answers that differ by 9.

> In general, we've
> $n^2 - p^2 = (n-p)(n+p)$

Answers
1. $n^2 - 4 = (n-2)(n+2)$
2. $n^2 - 9 = (n-3)(n+3)$

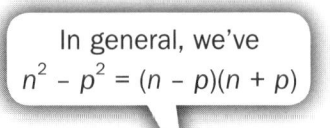

1 What is the remainder when the product of all the prime numbers between 1 and 1999 is divided by 10?

2 A *perfect square* is a number which is the square of an integer. How many integers greater than one million and less than 9 million are perfect squares?

3 What number must be added to the numerator and the denominator of the fraction $\frac{1}{2}$ to yield the fraction $\frac{2}{3}$?

4 What number is as much larger than 3×3 as it is smaller than 5×5?

5 Samuel wears socks of different colours. He has four pairs: one red, one blue, one yellow, and one green kept in the drawer. When he dresses, he removes a pair at random. How many days per week on average will he be seen wearing a matching pair?

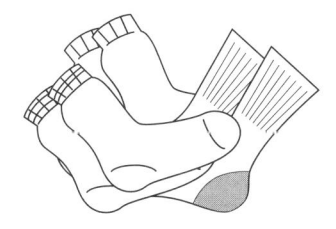

6 The first "*t*" of the recurring word *contestcontest*... occurs at the 4th letter of the pattern. Which letter of the pattern does the 777th "*t*" occur?

7 What is the least number of points one needs against one's opponent to win a "Best of three sets" match of tennis?

> 1 game = 4 points
> 1 set = 6 games

8 A triangle has sides 17 cm, 42 cm, and 59 cm. Find its area.

9. Solve the alphametic: $(\mathbf{HE})^2 = \mathbf{SHE}$, where each letter stands for a single digit.

10. How many zeros are there at the end of the product
$$1 \times 2 \times 3 \times \cdots \times 999 \times 1000?$$

> Nothing but ZERO
>
> K C Yan

5 More Mathematical Quickies & Trickies

Twitter Math @MathPlus

MathPlus is a 140-character microblog from *MathPlus Consultancy*. To subscribe to it, go to http://twitter.com/MathPlus

Are you a mathtwit? Get short, timely messages from Yan Kow Cheong. Join today and follow @MathPlus.

Breaking news in Singapore Math Education. Get your regular dose of math from K C Yan on Twitter.

Keep up with headlines and excerpts from the latest math news.
Something odd, often amusing, always informative.
New Math, Math News, Factoids, Tidbits, … and many more!

K C Yan tweets engaging stories, trivia, latest blog posts, math jokes, and upcoming events in the math community.

Follow him on Twitter via phones, web, or other application.

A Palindromic Equation? $2^5 - (5 + 2) = 25 = 5^2$ #math #mathjokes
11:44 AM Sep 6th via web

@mathguide For the Guinness Book of World Records, what is the most number of retweets recorded for a single tweet? #mathtweeory
September 7, 2010 1:03:55 AM via web

A googol and one is palindromic: 100…001 where "…" stands for 101_{1+1} 0's #mathfacts
1:58 AM Aug 21st via web

10 Daffynitions of a Calculator Singapore Math http://singaporemathplus.blogspot.com/?spref=tw
12:48 PM Feb 27th via web

What is the most likely blood group of mathematicians? And that of mathematics educators? (1) A (2) B (3) AB (4) O The answer is __. #math
12:08 AM Aug 11th via web

On average, there are 1.72 Friday the 13th's in any year. Pf: In a 400-year cycle, 146,097 days ÷ 688 Friday the 13th's = 212.35+ days #math
1:38 AM Aug 14th via web

Sine: The joker who gets the loan; Cosine: The elder who pays off the loan; Tangent: The banker who spends the money in Hawaii. #mathjokes
12:33 AM Aug 11th via web

Hypothesis: "There are more an atheist in mathematics than in science." A devilish proof: mATHEmatTIcS #mathjokes #math
2:10 AM Aug 5th via web

Murphy's law: When you dial the wrong number, it's never engaged! If you suspect it's imaginary, simply multiply it by i. #mathjokes #math
10:39 AM Aug 2nd via web

Ex-math teacher-turned-Khmer-Rouge-prison-chief, Duch, gets only 30 years' jail for the execution of 14,000-16,000 people. #math #Year Zero
12:12 AM Jul 28th via web

Why is −0 = 0? 0 is the additive identity: 0 + (−0) = −0 And, −0 is the additive inverse of 0: 0 + (−0) = 0. Thus, −0 = 0 + (−0) = 0. #math
4:02 PM Jul 19th via web

Happy "Birth Year"! You're born in 1955 & you're 55 in 2010. The year you reach the age equal to the last 2 digits of your birth year. #math
2:44 PM Jul 12th via web

ZEROLEXICON: Zeroklept = One who steals 'nothing'. Zerolater = One who worships 0. Zeromancer = One who practices divination by zeros. #math
1:22 PM Jul 8th via web

How many cat years equal one human year? 1-year-old cat ≈ 16 human years; 4-year-old ≈ a woman of 32; 8-year-old ≈ a 64-year-old #math
9:56 PM Jul 4th via web

Singapore may have the highest percent of cab driver licence holders in the world: 3 for every 100 Singaporeans! #mathfacts #singaporemath
12:08 PM Jun 24th via web

The lore of averages in traffic lights RED, RED, RED: I'm so unlucky! Zero luck GREEN, GREEN, GREEN: We take it for granted; that's not odd!
12:00 PM Jun 17th via web

On average, the rainbow is white. #mathjokes
12:56 AM Jun 16th via web

What is the parity of infinity? odd or even? ∞ is even if and only if it is odd! Proof: $\infty = \infty + 1$ #mathjokes
2:22 AM Jun 10th via web

Is Phi Divine or Demonic? $-2 \sin(666) \approx \varphi$ [Wang, Journal of Recreational Mathematics, V 26, 203] #Math
2:12 AM Jun 10th via web

Singapore Math: Let letters and numbers express themselves: ero 1ne 2wo 3hree 4our 5ive 6ix 7even 8ight 9ine 10n 11even #math
11:10 AM May 27th via web

The phone no. 6666666 was sold for $2.75 million—it beat the previous most expensive no. 8888888, sold a little over 1/2 of a million. #math
10:55 AM May 27th via web

Singapore Math: The fool hath said in his heart there is no empty set.
11:12 AM Apr 22nd via web

SIngapore MATH: "Including President Obama, 5 of the last 7 US presidents have been left-handed." And Singapore PM Lee is also left-handed.
March 7, 2010 12:10:38 AM via web

A prime number is one that can be divided not by 2 but 4 integers. 5 is prime: 5 is divisible by 1, 5, −1 and −5 (4 factors). A prime joke!
1:45 AM Apr 11th via web

"The Lord Bless you and Keep you." Numbers 6:24 "Fear not math, fear God." K C Yan
12:24 PM Feb 25th via web

For more mathtwits, follow K C Yan on Twitter as @MathPlus

1. As I was going to St. Eves, I met a man with 7 wives. Every wife had 7 sacks, and every sack had 7 cats, every cat had 7 kitten. Kits, cats, sacks and wives, how many were going to St. Eves?

2. Write down ten million ten thousand ten hundred and ten.

3. A seqeunce of numbers had its first term equal to $\frac{3}{4}$. Each new term is obtained by the formula $\frac{1-x}{2+x}$, where x is the previous terms. What is the 100th term?

4 What is the remainder when 2^{2002} is divided by 10?

5 If $x + y = 1$ and $x^2 + y^2 = 2$, what is the value of $x^4 + y^4$?

6 When Samuel and David work together, they can complete a piece of job in 7 hours. If David does twice as much work as Samuel, how long would David take to do the job by himself?

7. A woman born in the 18th century was x years old in the year x^2. How old was she in 1756?

8. If $p + q = 1$, where p and q are positive numbers, find the smallest possible value of $\frac{1}{p} + \frac{1}{q}$.

9 If the side of a square increases by 25%, what is the percentage increase of the area of the square?

10 A train 1 km long is travelling at a steady speed of 30 km/h. It enters a tunnel 1 km long at 2:00 pm. At what time does the rear of the train come out of the tunnel?

More Mathematical Quickies & Trickies

What Is 27 × 37, Really?

Let's look at a number of ways how to mentally multiply two numbers.

Method 1

$27 \times 37 = (30 - 3) \times 37$
$= 30 \times 37 - 3 \times 37$
$= 1110 - 111$
$= 1110 - 110 - 1$
$= 1000 - 1$
$= 999$

Method 2

$37 \times 27 = (40 - 3) \times 27$
$= 40 \times 27 - 3 \times 27$
$= 1080 - 81$
$= 1080 - 80 - 1$
$= 1000 - 1$
$= 999$

Method 3

$27 \times 37 = (25 + 2) \times 37$
$= 25 \times 37 + 2 \times 37$
$= \frac{3700}{4} + 74$
$= 925 + 74$
$= 925 + 75 - 1$
$= 1000 - 1$
$= 999$

Method 4

$27 \times 37 = 3 \times 3 \times \underbrace{3 \times 37}$
$= 3 \times 3 \times \underbrace{111}$
$= 3 \times 333$
$= 999$

Method 5

$27 \times 37 = (32 - 5)(32 + 5)$ $\quad a^2 - b^2 = (a - b)(a + b)$
$= 32^2 - 5^2$
$= 1024 - 25$
$= 1024 - 24 - 1$
$= 1000 - 1$
$= 999$

> Think of 2 more ways of solving 27 × 37.

Practice

Use the difference of two squares to evaluate the following.

(a) 29×31 (b) 37×43 (c) $5\frac{1}{2} \times 6\frac{1}{2}$ (d) $8\frac{1}{4} \times 7\frac{3}{4}$

Answers

(a) $29 \times 31 = (30 - 1)(30 + 1)$
$= 30^2 - 1^2$
$= 900 - 1$
$= 899$

(b) $37 \times 43 = (40 - 3)(40 + 3)$
$= 40^2 - 3^2$
$= 1600 - 9$
$= 1591$

(c) $5\frac{1}{2} \times 6\frac{1}{2} = \left(6 - \frac{1}{2}\right)\left(6 + \frac{1}{2}\right)$
$= 6^2 - \left(\frac{1}{2}\right)^2$
$= 36 - \frac{1}{4}$
$= 35\frac{3}{4}$

(d) $8\frac{1}{4} \times 7\frac{3}{4} = 7\frac{3}{4} \times 8\frac{1}{4}$
$= \left(8 - \frac{1}{4}\right)\left(8 + \frac{1}{4}\right)$
$= 8^2 - \left(\frac{1}{4}\right)^2$
$= 64 - \frac{1}{16}$
$= 63\frac{15}{16}$

> Creative mental calculation may prevent premature dementia (or memory loss).

1 What 2-digit number exceeds its reversal by 20%?

---

2 The smallest angle in a triangle is 20°.
Find the largest possible angle in the triangle.

An acute angle is a cute angle!

3 The Bible Society consists of 1243 groups of equal sizes. The members can also be grouped into 23,843 town cells. What is the least number of members of which the society can be made up of?

Did you know...
The Bible
(Holy Scriptures)
is often known as the
"Good News."

Bad news: 9/11, 26/11
Good news: Numbers 6:24

4 John and Jane have some money. If John gives Jane $5, they will have the same amount of money. If Jane gives John $5, John will have twice as much money as Jane. How much does each have?

38 More Mathematical Quickies & Trickies 6

5. A motorboat traveled upstream in 3 h, but took only 2 h for the return downstream trip. If the current flowed at 8 km/h, find the speed of the boat in still water.

6. A tank can be filled by one pipe in one hour, by another in two, by a third in three, and by a fourth in four. How long will it take for all four pipes together to fill the tank?

7 How many rectangles are there in an 8×8 chessboard?

8 Given that two people out of every five in a school like geometry. If a sample is taken, what is the chance (or probability) that two people in the sample like geometry?

9 Show that $\sqrt{1+\sqrt{1+\sqrt{1+\sqrt{1+\ldots}}}} = \dfrac{1}{\sqrt{1-\sqrt{1-\sqrt{1-\sqrt{1-\ldots}}}}}$.

10 Solve the simultaneous equations: $2x + y + 2z = 14$
$x + 2y + z = 16$
$x + y + 2z = 18$

7 More Mathematical Quickies & Trickies

Humanizing 1, 2, 3

Use the digits 1, 2, 3 to creatively quantify some feelings and emotions. Here is an off-the-wall sample of mine. Share yours with the rest of us.

FEAR: 2, 1, 3, 1, 3, 2, __, __

FAITH: 1, 2, 3, ...

HOPE: 123, 132, 213, 231, ...

HEALTH: 12, 21, 32, ..., 112?

WEALTH: $\$3^{12}$

JOY: 1–2–3, 1—2—3, ...

STRESS: Last two digits of 3^{21} = ?

PERSEVERANCE: 2nd, 3rd, 1st

PERFECTION: 2.31 s, 2.13 s, 1.32 s

FEEDBACK: 1, 2, 3, 1, 2, 3, ...

FOCUS: 1-2-3, 1-2-3, ...

OPTIMISM: 12, 21, 23, 32, ...

HUMOR: 1ne, 2wo, 3hree

GOAL: 1 + 2 = 12

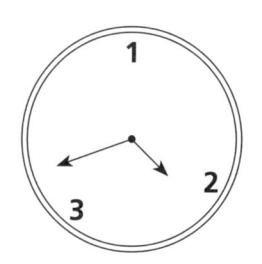

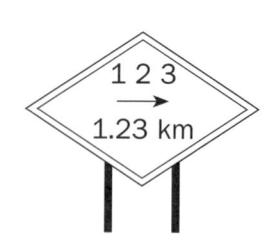

BEAUTY: $12 \times 13 = 156$; $21 \times 3 = 651$

SYMMETRY: $121 \times 11 = 1331$

COINCIDENCE: $1 + 2 + 3 = 1 \times 2 \times 3$

PROBLEM SOLVING: 123! ends with ? zeros.

PROBLEM POSING: $n!$ **ends with 123 zeros. Find** n**.**

NUMEROLOGY: $12 = 3$

PRAYER: $1 + 2 \neq 3$

SURPRISE: $1, 2, 3, \pi$

LIE: $2 > 3$

FUTURE: DAY 1 + DAY 2 + DAY 3 + ...

FUN: $1^{2^3} = 1^{2^3}$

GOSSIP: $1 + 2 = 3 \pm \varepsilon$

HUMILITY: $\frac{12}{13} > \frac{21}{31}$

TRUTH: $\frac{1}{2^1} + \frac{1}{2^2} + \frac{1}{2^3} + \cdots = 1$

MIRACLE: $1 = 2 = 3 = \cdots$

UNMERITED FAVOR: $f(\dagger) = \frac{\text{Health}}{(123)} + \frac{\text{Wealth}}{(\$2^{31})}$

> As *Difficult* as
> **1, 2, 3**
>
> K C Yan
>
> 2011

Humanize your own version with 1, 2, 3!

1 Siti and her friends stand in a circle. Both neighbors of each child are of the same gender. If there are five boys in the circle, how many girls are there?

2 How many squares are there in an 8 × 8 chessboard?

> A square is a rectangle, but a rectangle is not a square.

3 Given that $x + y = 2$ and $xy = 3$, find
  (i) $\dfrac{1}{x} + \dfrac{1}{y}$,   (ii) $\dfrac{1}{x^2} + \dfrac{1}{y^2}$,   (iii) $\dfrac{1}{x^3} + \dfrac{1}{y^3}$.

4 What 2-digit number is the square of its units digit?

$ab = b^2$

5 If Sally can type one digit in one second, how long will it take her to type out all the numbers from 1 to 2000?

6 Mr. Yan said: "The day before yesterday I was 40, but I will turn 43 in the next year." How could this be possible?

7 What day follows the day before yesterday if 3 days from now will be Wednesday?

8) Given that $n^2 + 1$ contestants sign up for a tennis tournament. How many games must be played to determine the winner?

9) The diameter of the inner circle is 4 cm and of the outer circle is 6 cm. Which is smaller: the area of the inside circle, or the area between them?

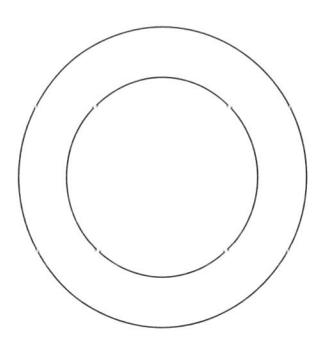

10) Which is larger: 2^{30} or 3^{20}?

More Mathematical Quickies & Trickies 7

8 More Mathematical Quickies & Trickies

A Mathophobia Kit

You may be suffering from some degree of mathophobia if you experience some of these:

- [ ] You break into a sweat on hearing the consumer price index (CPI) being discussed at a party, or on having to figure out the tip in a restaurant.

- [ ] Your heartbeat increases because you cannot find a calculator to do some simple multiplication and division.

- [ ] Your mind goes blank because you need to read a statistics module for your diploma course.

- [ ] You dread the math section of the ACT, SAT or GMAT paper.

- [ ] You skip any questions on probability, and permutations and combinations when you have a choice.

- [ ] You memorize the solutions, instead of trying to understand the principles behind the questions.

- [ ] You opt for a non-mathematical module when there is a choice.

- [ ] You give any programming course a miss lest you may not follow the logical steps in writing a program.

- [ ] You switch to paying by credit card, instead of paying by cash even though the latter can save you some tens of dollars.

- [ ] You have nightmares in having to solve word problems like, "If it takes John 3 hours to mow a lawn and Jane 2 hours to mow the same lawn, how long will it take if they mow it together?".

- [ ] You would rather do household chores than help your child with some word problems.

- [ ] You frequently dream of being in an examination hall, sitting for a math paper; or dread of not having enough time to prepare for your math paper.

☐ You delegate the task of filling your income tax returns form to your spouse, or pay someone to do it for you.

☐ You skip the financial pages of your paper, or switch channels when financial news is on TV.

☐ You avoid superstitious or unlucky numbers lest you fall victim to some unfortunate events.

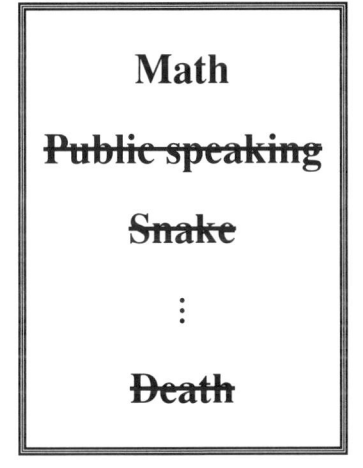

Math

~~Public speaking~~

~~Snake~~

⋮

~~Death~~

The No. 1 Fear in the world isn't death or ghosts, but the language of science and technology!

Mathophobia Discrimination Bill

It is illegal to discriminate against anyone with numerical symptoms.

Is mathophobia curable?

1 The sum of two numbers is 26, and their product is 153. What are the numbers?

2 A locker has developed a four-digit code that is unbreakable, except if the code begins with 0, 2 or 5. What is the greatest number of four-digit codes the locker can use that will not be broken?

3 The year 1991 is divisible by 11, and the year 1992 is divisible by 12. When will we have a year in the 22nd century that is divisible by both 11 and 12?

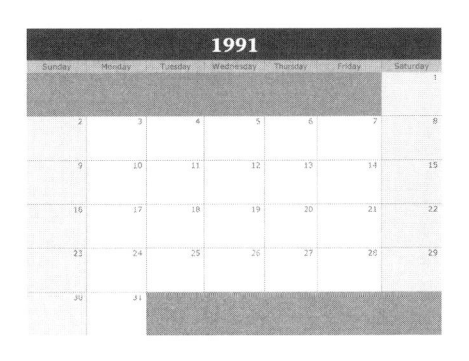

4 In Euclid's Homes, there are 16 houses. How many different routes can you take from point X to point Y, moving only upwards and to the right?

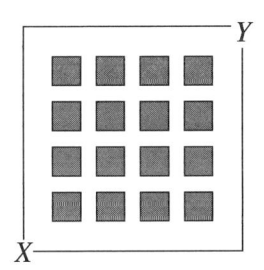

5. On Sunday Mrs. Singh bought 16 m of cloth. On the next day, Monday, she cut off 2 m. In the next few days, she cut off 2 m of material every day, except on the following Sunday, when she went to church. On which working day did she make her last cut?

6. Place the numbers 1 through 9 in the boxes below so that both multiplications are true.

☐☐ × ☐ = ☐☐☐ = ☐☐ × ☐

52   More Mathematical Quickies & Trickies 8

7 A set of traffic lights runs through the following sequence: red (90 s), red and amber (5 s), green (80 s), amber (5 s), then back to red, and so on. How long is the red light on during a 24-hour period?

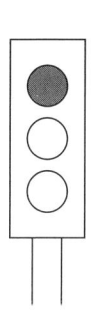

8 Joe walks to school every day. Every morning he leaves home and reaches the main road using one of the six paths, as shown below. He then travels on part of the one-way road and gets back home by another path. How many different routes can Joe take?

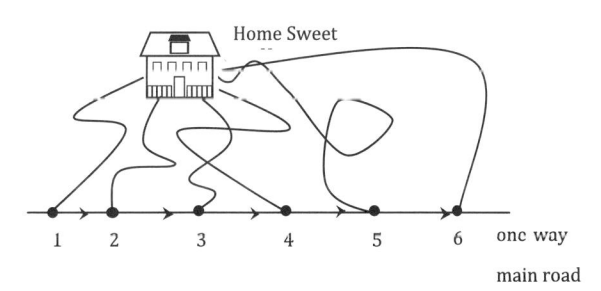

9. In how many ways can the word MATHEMATICS be spelled starting from the top and working down through the array if
   (a) we can select any letter from each row,
   (b) we can only select either one of the two neighboring letters directly underneath?

```
                M
              A   A
            T   T   T
          H   H   H   H
        E   E   E   E   E
      M   M   M   M   M   M
        A   A   A   A   A
          T   T   T   T
            I   I   I
              C   C
                S
```

10. A number formed by the digit 1, 2, 3, 4, 5, 6, 7, 8, 9 is such that the number formed by the first n digits is divisible by n for all values of n from 1 to 9. What is the number?

Test your divisibility tests!

More Mathematical Quickies & Trickies

WITs — 13 Ways to Attain Mathematical Excellence

1. **Have all certificates carry an expiry date.**

2. Encourage more mathematicians and mathematics educators to take up politics.

3. **Encourage more girls to have faith in mathematics.**

4. Eradicate innumeracy by exposing pseudo-mathematics (e.g., Numerology, Horoscope, *I Ching*, ...).

5. **Denounce political and bureaucratic interference for concealing mathematical illiteracy among certain groups in the population.**

6. Encourage competence in rounding off measurements and large-scale estimates.

7. **Construct a mathsemantic web to better handle problems involving both math and semantics — e.g., *What is the sum of 2 apples + 3 oranges?* and *What is the result of 3 round trips + 2 half trips?*.**

8. Commemorate Math Days such as *Statistics Day, Pi Day* (3/14), *CHRISTmaths Day*, and the like.

9. **Expose students to fallacies and paradoxes to cleanse their minds of mathematical loopholes and myths.**

10. Expose statistical massage from politicians and unscrupulous advertisers.

11. **Shy away from teachers who have taught Math for over 10 years, but have only one year of teaching experience repeated over ten times.**

12. Flee from millionaires-tutors who are charging their students exorbitantly.

13. **Encourage teachers wishing to continue teaching Math after a decade to sit for a math upgrading proficiency test.**

1 Mentally compute the following.
   (a) $101^2 - 201$
   (b) $99^2 + 200 - 1$

2 Sam gets $3 more pocket money than Bob each week. They each spend $15 weekly on food and save the rest. When Sam saves $72, Bob only saves $48. How much pocket money does Bob get each week?

3. Just before he dozed off, Joe remembered that the product of the pages of his Algebra book was 23,870. What were the page numbers?

4. The ratio of the number of $2 notes to $10 notes in Paul's pocket was 1 : 5. When Paul exchanged 20 pieces of $10 notes for $2 notes, he then had an equal number of $2 and $10 notes. How much money did Paul have in his pocket?

> I'm called nix, nada, zip, zilch, zippo, ... and I've magical powers to dissolve you to nothing.

0

5 A troop has enough food to last for 24 days. By how much will each recruit's ration have to be reduced if the food is to last 40 days?

6 On a map of scale 1 : 200,000, the area of a region is 6 cm^2. Calculate
  (a) the actual area, in km^2, of the region,
  (b) the area, in cm^2, representing the region on a second map of scale 1 : 20,000.

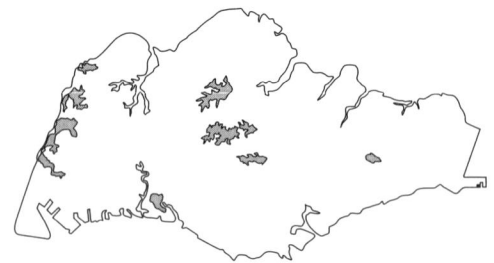

7 Jane is pasting stickers inside an album. If 1 sticker is pasted on each page, 4 stickers will be left over, and if 2 stickers are pasted on each page, 1 page will remain empty. How many stickers and how many pages are there in the album?

8 An egg seller sells half of all the eggs she has plus half an egg to her first customer. She then sells half of the remainder plus half an egg to her second customer, and half of the remainder plus half an egg to her third customer. She repeats the same process for the next customer and so on. After she has served her seventh customer, she finds that she has sold all her eggs. How many eggs did she have originally?

9. If a factory worker makes 240 toys a day, 400 fewer toys will be made. If she makes 280 toys a day, 200 more will be made. How many toys should the worker make, and how many days has been set for it?

10. Two boys are at points A and B, respectively. The first travels the distance between A and B in x hours, the second in y hours. In how many hours will the boys meet if they leave points A and B at the same time in order to meet each other?

10 More Mathematical Quickies & Trickies

Facebook Math: Numeracy vs. Literacy

Numeracy vs. Literacy
by Yan Kow Cheong on Tuesday, September 7, 2010 at 11:50pm

Which is more important: Literacy or Numeracy? Why is it considered socially acceptable to say, 'I'm no good at math?' People announce they are no good at math in a manner that suggests they are actually proud of it. Yet, you'd hardly hear people boasting about their linguistic ineptitude.

Does numeracy (or *quantitative literacy*, as it's called in the United States) matter less than literacy? The penalties for the illiterates seem to far outweigh those of being innumerate. You wouldn't be able to read a newspaper, fill in a form, order from a menu, or follow road signs, or read a map. Or, you wouldn't be able to operate your iPod, use the Internet, or gossip on Facebook.

In reality, being numerate has a profound impact on how successful you are, far greater, in fact, than being literate.

Both literacy and numeracy have a bearing on earnings; however, numeracy seems to have a more powerful effect than literacy. Low earnings is almost guaranteed if one has poor basic skills than one has good basic skills, but the difference in earnings is greater for numeracy than for literacy.

At the national level, numeracy has a profound effect on the average productivity of the workforce and explains a significant proportion of the difference in economic performance between nations.

Good businesses go bust because the MD couldn't get his 'head around the figures.' Bosses with average quantitative skills would find it hard figuring out the metrics of different components of their businesses; as a result, many conveniently leave the job to creative accountants or unscrupulous managers to doctor the figures to keep the shareholders happy.

Singapore's footballer Fandi Ahmad, ex-professional boxer Mike Tyson, and pop music entertainer Michael Jackson, to name a few, were taken advantage of financially, probably due to their below-average numeracy skills.

Matt and Luke Goss of the band Bros lost an entire fortune because they didn't understand the difference between *gross profit* (the amount of money you receive in a single year) and *net profit* (the amount of money you've left once you've deducted all your costs.)

Here are some words from the mouths of sportsmen:

David Beckham: "The team would be giving 110 per cent in the match against Azerbaijan."

Kevin Keegan: "Nichol never gives more than 120 per cent."
Kevin Keegan often demanded 1000 percent when he was manager of Newcastle.

Steve Davis, on a snooker match: "It was a game of three halves."

Here are more numerical dim wits:

I double-checked it six times." Maria Berra

"Cut the pizza into four slices, not eight. I'm not hungry enough to eat eight." Yogi Berra

"If you believe in God, you'll believe in nothing." Clair Patterson

How old is planet Earth? 4.55 billion years
Less than 10 000 years old according to Creationists

"Never go for a 50/50 ball unless you are 80/20 sure." Ian Darke

"She's 80 per cent nuts, 20 per cent normal, 95 per cent talented"
                    Simon Cowell, on an *X Factor* contestant

Visitor: "How does Singapore Airlines stay on top all these years?"
Sim Kay Wee, SIA Senior Vice President: "100% is not enough. When you reach #1, you need 120%."

Of course, you can argue that these fellows are already well-off, even though they're probably numerically challenged, but they'd be financially better off if they're less innumerate.

Many of our parents and relatives with little formal education go through life relatively unscathed (and some even did pretty well) because they're basically numerate rather than literate. Imagine a hawker or shop owner who can't handle basic math, or is allergic to handling money. Or, even from the criminal or dark world, can a *loan shark* or drug lord be mathophobic? Their real-life examples show that although literacy matters, however, numeracy matters more!

So, let's stop flirting with innumeracy, and take steps to sign up for some INTERMEDIATE or ADVANCED NUMERACY to be mathematically street-smart.

Numerically yours

KC Yan
@MathPlus
http://singaporemathplus.blogspot.com/

[The above five minute mathematics article was originally posted on Facebook.]

1. Without the use of calculators, find the following.
   (a) $300^2 - 301 \times 299$
   (b) $999^2 + 1999$

2. Find the next integer after 1 that is simultaneously a perfect square, a perfect cube, a perfect fourth power, a perfect fifth power, ..., and a perfect ninth power.

3. When an integer n is divided by 2747, the remainder is 79. What is the remainder when n is divided by 67?

4. Two trains 20 km apart travel towards with each other at 40 km/h. A fly which always travels at 60 km/h starts off on the first train and flies towards the second. When she gets to the second train she turns around and heads straight for the first. She continues this journey until she is squashed in the head on collision. How far does the fly travel?

5. It is 1:00 a.m. on a Sunday in Hong Kong, China. What time would it be in these other cities?
   (a) Sydney, Australia
   (b) Bengalore, India
   (c) Singapore, Singapore
   (d) Moscow, Russia
   (e) Tokyo, Japan

6. A bottle $\frac{3}{4}$ full of a liquid weighs $\frac{3}{2}$ kg. The liquid alone weighs $\frac{3}{4}$ kg more than the empty bottle. Find the weight of the bottle when it is full.

7 A boy rowed 6 km upstream in $1\frac{1}{2}$ h and 6 km downstream in $\frac{3}{4}$ h. Calculate
(a) the speed of the current,
(b) the speed of the boat in still water.

8 Job and Ruth have savings in the ratio 3 : 4. Ruth saves another 20% of her saving. How many percents of Job's money must he save so that he would have twice as much as Ruth in total?

9 Two cylists are moving in the same direction on a circular track whose circumference is 999 m, and they meet every 37 minutes. Given that the speed of the first is 4 times greater than the speed of the second, find the speed of each cyclist.

10 Divide the number 100 into 4 unequal parts so that if 4 is subtract from the first number, 4 is added to the second number, 4 is multiplied to the third number, and the fourth number is divided by 4, the same result is obtained in all four cases. What are the four numbers?

11 More Mathematical Quickies & Trickies

Thou Shalt Not Divide By Zero

Why can't we divide by zero? The answer involves the idea of consistency. Division by zero leads to either *no* number or to *any* number.

To divide a by b means we need to find a number x such that $bx = a$, thus $x = \frac{a}{b}$.

If $b = 0$, there are two different cases to discuss (i.e. $a \neq 0$ and $a = 0$).

Case 1: $a \neq 0$

Since $a \neq 0$ and $b = 0$, $x = \frac{a}{b} = \frac{a}{0}$, or $0 \times x = a$.

What number x, multiplied by 0, will yield a, where a is any fixed number, not equal to 0?

Since any number multiplied by 0 is 0, there is no such number x.

Case 2: $a = 0$

Since $a = 0$ and $b = 0$, $x = \frac{a}{b} = \frac{0}{0}$, or $0 \times x = 0$

Because any number multiplied by zero is zero, x can take any number.

Therefore, division by zero leads either to *no* number or to *any* number.

Zero is really a troublemaker!

Alternatively,

Why is $\frac{1}{0}$ an illegal operation?

Let us assume that $\frac{1}{0}$ is allowed and is equal to some number.

By the rules of arithmetic, $0 \times$ (any number) $= 0$

So, $0 \times \frac{1}{0} = 0$ \qquad (1)

Another rule says that: $x \times \frac{1}{x} = 1$, provided x is not zero.

But, if we want $\frac{1}{0}$ to be a number, then, we have $0 \times \frac{1}{0} = 1$ \qquad (2)

By equations (1) and (2), $1 = 0$. A contradiction!

So the assumption that $\frac{1}{0}$ exists is WRONG.

"Thou shalt not divide by zero" is known as the Mathematician's *11th commandment*.

God is a Great Mathematician!

1 If $a + b + c = 1$, prove that $ab + bc + ac < \frac{1}{2}$.

---

2 If x and y are real numbers, find the smallest possible value of
$$x^2 + y^2 - 8x + 6y + 17.$$

3 Esau and Jacob take part in a race. Each boy accelerates at a uniform rate from the start. Esau covers the last quarter of the distance in 3 seconds and Jacob covers the last third in 4 seconds. Who won the race?

4 A speedometer reads 10% more than the car's traveling speed. How fast is the car actually traveling if the meter reads 100 km/h?

5 An integer is *decreasing* if each digit is less than the one to its left. For example, 5320 is decreasing. How many decreasing integers occur between 2000 and 7000? (*Mathematics Teacher, Vol. 90, No. 6, Sept. 1997*)

6 If Mark walks up an upgoing escalator at the rate of one step per second, 20 steps bring him to the top. If he walks up at two steps per second, he reaches the top in 32 steps. How many steps are there in the escalator?

Be a *Mathepreneur*, today!

1. Design a 25-hour clock

2. Write a math bestseller.

3. Invent an electrified abacus.

⋮

10. ...

The Little Book of
BIG MATH IDEAS

7 Given two pieces of fuse, each of which burns in 1 minute. Without scissors, and noting that each fuse may have a variable rate of burning, how would you time 45 seconds with these pieces of fuse?

---

8 Express $x^4 + 4$ as the difference of two squares.

9 In a Chinese village, every husband and wife keep having children until they have a boy and then stop. Assuming boys and girls are equally likely, will this produce more baby girls or more baby boys in the whole village?

10 Suppose that the minute and hour hands on a clock were identical. When would it be impossible to tell the time?

> **Math is the ANSWER!**
>
> 1. Why aeroplanes can fly.
> 2. Why on-line purchase is safe.
> 3. Why tigers are stripy but leopards are spotty.
>   ⋮
> 10. Why black holes exist.

12 More Mathematical Quickies & Trickies

Math Jokes to Relieve Stress

Q: What did the 0 say to the number 8?
A: 'Nice belt!'

Q: **Why was six scared of seven?**
A: **Because seven eight nine.**

Q: Why is nine drunk?
A: Because it is one over the eight.

Q: **What kind of ant is good at math?**
A: **An accountant.**

Subtraction

Teacher : If you had nine dollars and I took away four, what would you have?
Student : A fight.

Fraction/Division

Teacher : If your dad earned $1000 a week and gave you half, what would you have?
Student : A heart attack.

Teacher: If 12 eggs cost $1.20, how many would you get for 60¢?
Student: None.
Teacher : None?
Student : None. If I had 60¢, I will get some candies.

Q: There were ten zebras in the zoo until all but nine escaped. How many were left?
A: Nine.

Q: Why 2 × 10 = 2 × 11?
A: Twice ten is twenty, and twice eleven is twenty too.

Q: Why was the math book sad?
A: Because it had so many problems.

Or the book has no answers to check!

Q: Why is long division such hard work?
A: Because of all the numbers you have to carry.

Q: Which is correct?
'nine and five is thirteen'
or
'nine and five are thirteen'
A: Neither. Nine and five are fourteen.

Q: Why did the math coach take a ruler to bed with him?
A: To see how long he would sleep.

Q: Did you hear about the algebra teacher who fainted in class?
A: Everyone tried to bring her two.

Q: What meals do math teachers like to eat?
A: Take aways.

Q: What do math tutors like for dessert?
A: Pi.

1 Which is greater: 2^{40} or 3^{30}?

2 A seller remembers that when he counted his eggs by twos, threes, fours and fives, he had remainders of 1, 2, 3 and 4 eggs, respectively. How many eggs did he have?

3 Suppose a man has two children in the year 2000. If each of the children has two children 25 years later, and each of those children has two children 25 years later, and so on. How many new descendants are born in the year 3000?

4 How many times does the minute hand pass the hour hand between 12 noon and 12 midnight?

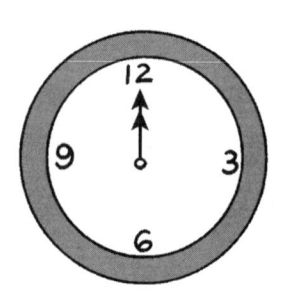

5 Factor $x^3 - y^3 - z^3 - 3x^2z + 3xz^2$.

6 Ruth spent exactly $2 on stamps. She bought some 4¢ stamps, ten times as many 2¢ stamps and made up the balance with 10¢ stamps. How many stamps of each value did she buy?

7 Show that $0.249999\ldots = \frac{1}{4}$.

8 (a) What is the last digit of 7^{1000}?
(b) Find the units digit of 27^{27}.
(c) Find the sum of digits of $(10^{1998} - 1)$.

9 A cow costs $1000, a pig $300 and a sheep $50. Farmer Yan buys 100 animals and at least one animal of each kind, spending a total of $10,000. How many of each kind did he buy?

10 At 12 noon, the hour hand, the minute hand, and the second hand, of a clock all coincide. Is there another time, before it is 12 o'clock again, when all three hands are exactly together?

13 More Mathematical Quickies & Trickies

Look-see Proofs

Look at the three wholes, each one being divided into quarters. Can you see why 3 divided by $\frac{1}{4}$ is 12?

$3 \div \frac{1}{4} = 12.$

Use a geometric argument to determine the number of three-quarters in three and three-quarters.

A Visual Proof that $3 \times 6 = 6 \times 3$

Let us interpret 3×6 geometrically.
"Three times six" means three groups of six things.

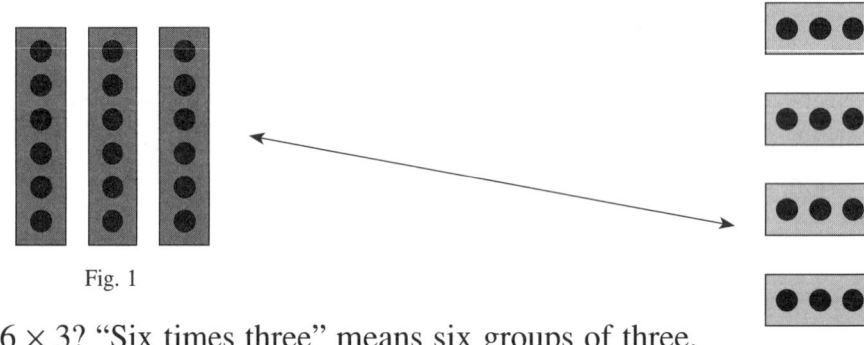

Fig. 1

What about 6×3? "Six times three" means six groups of three.

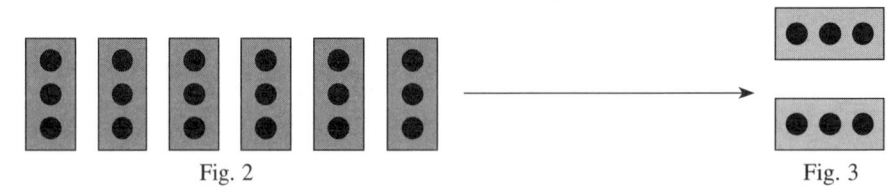

Fig. 2   Fig. 3

Figure 1 and Figure 2 are quite different.

However, we can see a connection between the two figures when we group the dots in Figure 2 into those depicted in Figure 3.

Figures 2 and 3 show why $3 \times 6 = 6 \times 3$.

Seeing is believing!

Use a "visual method" to solve the following questions.

1. **The Singapore Model Method**

    The sum of two numbers is 13.
    Their difference is 5.
    What are the two numbers?

2. **Visualizing Multiplication**

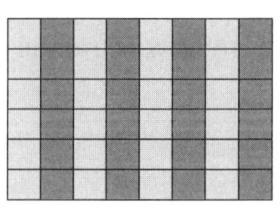

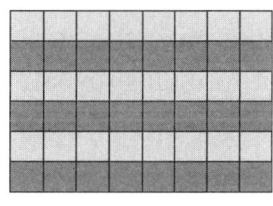

    ? × ? = ☐   ? × ? = ☐

3. **The sum of the interior angles in a triangle is 180°.**

4. **Sum to infinity**

$$\tfrac{1}{2} + \tfrac{1}{4} + \tfrac{1}{8} + \tfrac{1}{16} + \ldots = ?$$

5. **Different Geometrical Interpretations**

$$A = \pi r^2 \qquad A = \tfrac{\pi}{4} d^2 \qquad A = \tfrac{1}{2} cr$$

The first formula gives the area of a circle with radius r. What do the other two equivalent formulas reveal?

6. **Why is 12 + 1 = 11 + 2?**

Think in words!

7. A visual proof of 15 × 13 = 195

8. Why is $a - (b + c) = a - b - c$?

1 Find the next number in the sequence: 2, 4, 8, 16, 31, ___.

2 Aaron can run around a track in 40 seconds. Betty, running in the opposite direction, meets Aaron every 15 seconds. How long does Betty take to run around the track?

3 What is the units digit for the sum $1^3 - 2^3 + 3^3 - 4^3 + 5^3 - 6^3 + \cdots + 19^3$?

4 John has walked $\frac{2}{3}$ of the distance across a railway bridge when he sees a train approaching, at 45 km/h. If John can just escape by running at uniform speed to either end of the bridge, how fast must he run to avoid the train?

5 The length and width of a rectangle are x cm and y cm, respectively, where x and y are integers. If $xy + y = y^2 + 13$, find the maximum area of the rectangle.

6 If y and $\frac{667}{y}$ are both integers, find the total number of possible values of y.

7. The perimeter of a rectangle is $16\sqrt{2}$ cm. Find the smallest possible value of the diagonal.

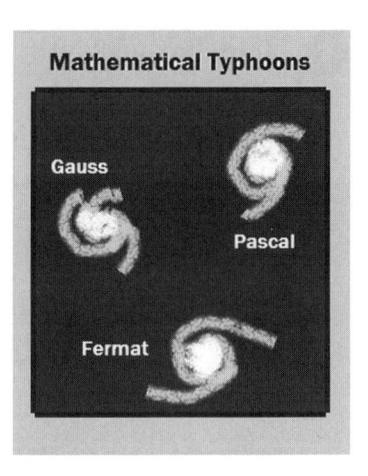

Mathematical Typhoons

Gauss

Pascal

Fermat

8. A solution to the equation $(x + a)(x + b)(x + c) + 5 = 0$, where a, b and c are different integers, is $x = 1$. Find the value of $a + b + c$.

9 Let p and q be positive integers. If p is a perfect square and the difference between $(p + q)$ and pq is 1000, find q.

10 A group of 43 Rotarians spent $229 at a breakfast buffet. Each man paid $10, each woman paid $5 and each child paid $2. What was the largest possible number of male Rotarians in the group?

14 More Mathematical Quickies & Trickies

Some PhD Math Questions

Other than those preparing for a PhD in mathematics, even mathematics majors may have little idea or exposure to the kind of questions graduate students have to answer to get a place into a PhD mathematics programme. Below is a sample of some higher arithmetic questions, which were set in some PhD qualifying entry examinations.

1. **Determine the rightmost decimal digit of $A = 17^{17^{17}}$.**

2. Determine the last digit of $23^{23^{23^{23}}}$ in the decimal system.

3. **Suppose that $n > 1$ is an integer. Prove that the sum $1 + \frac{1}{2} + \cdots + \frac{1}{n}$ is not an integer.**

4. For three nonzero integers a, b, c, show that
$$\text{HCF}\,[a, \text{LCM}\,(b, c)] = \text{LCM}\,[\text{HCF}\,(a, b), \text{HCF}\,(a, c)]$$

5. **Prove that the number π is irrational.**

6. Find all pairs of integer a and b satisfying $0 < a < b$ and $a^b = b^a$.

7. **Which rational number t are such that**
$$3t^3 + 10t^2 - 3t$$
   is an integer?

8. Evaluate the limit
$$\cos\frac{\pi}{2^2} \cos\frac{\pi}{2^3} \ldots \cos\frac{\pi}{2^n}.$$

Most of these post-graduate mathematics questions are easy to understand, and bright high school students may even comprehend their solutions; however, the thinking processes to arrive at the answers are far from obvious—maybe light years away even for the educated public.

1 Find the value of $1999 \times 19981998 - 1998 \times 1991998$.

2 What is the next number in this series?

$$1 \quad 4 \quad 2 \quad 8 \quad 5 \quad 7$$

> Q: Divide the circumference of the moon by its diameter.
> A: Pi in the sky.

3 Find the value of $1^2 - 2^2 + 3^2 - 4^2 + \ldots + 1999^2 - 2000^2$.

4. Mr. Ian's coin collection is kept in three bags. One fifth of the coins are in the first bag, several sevenths in the second bag, and there are 303 coins in the third bag. How many coins does Mr. Ian have?

5. A train left town P to town Q. At the same time, another train left from town Q to town P in an opposite direction. After they met, one train needed another nine hours to reach Q and the other train needed another four hours to reach P. How long will the slower train take to travel from P to Q?

6 If a and b are positive integers such that $a - b = 75$ and LCM $(a, b) = 360$, find the value of $a + b$.

7 If a and b are positive integers such that $a - b = 165$ and HCF $(a, b) = 15$, find the value of $a + b$.

8 If 7^{2009} is divided by 10, what will the remainder be?

Q: If 666 is the number of the beast, what is 670?
A: The approximate number of the beast.

9 Given that the sum of the infinite series
$$\tfrac{1}{2} + \tfrac{1}{4} + \tfrac{1}{8} + \tfrac{1}{16} + \cdots = 1$$
what is the sum of the infinite series
$$\tfrac{1}{4} + \tfrac{1}{16} + \tfrac{1}{64} + \tfrac{1}{256} + \cdots ?$$

---

10 In a class, 52% of all students are movie addicts, and 25% of them are book lovers. If a student is selected at random, when is the chance (or probability) that he or she is not a movie addict and is not a book lover?

15 More Mathematical Quickies & Trickies

Mathematical Prayers

O holy St. Gabriel, who prayed that in your algebra paper you would be asked only the questions you know, grant now that I too may be asked only the questions I know.

Heavenly Father, help me overcome my phobias, especially my mathophobia. Also, since Friday the 13th is round the corner, help me also overcome my *triskaidekaphobia*. The sight and sound of numbers paralyze me every time. My heart beats like a woman giving birth; my palms become terribly sweaty, and my mind becomes blank suddenly. Yes Father God, heal my numerical symptoms.

Lord Jesus, you're lucky there was no PSLE during your time. I want to do well for my GCE O-Level Math so that I'd become a famous mathematician one day, and maybe win a "Nobel Prize in Mathematics" when I grow up!

Dear God, you know I couldn't choose the project I want for my dissertation. I pray that my friends will have demanding supervisors, that they will eventually drop out, or switch to other topics of lesser importance.

O Lord, you know I didn't practice my math problems. If it's Your Will that I deserve to fail, I accept Your Will, Father, but I pray that my friends won't do well either.

God, help me to score a Distinction in my PSLE Math. Also I want to show Mr. Lim that I can get better than 98 for my paper. Help me, Lord. I promise that I will teach my brother in Math if I get an A^+ for my Math paper.

Father, there was E-Maths and A-Maths in secondary school. Now I've to prepare myself for the C-Maths and F-Maths, Lord. Help me to F-A-C-E Maths confidently.

God, some of my friends are hoping for an "A" in their Math paper. The majority are aiming for a good pass. You know, Lord, I'll be very happy if I can score an "E" for Math. I pray that the paper will be difficult so that everybody won't do well in it. Then the Cambridge examiners would have to shift the bell curve so that even the weak ones like me can make it. In fact, God, the harder the paper is, the better it is for me. Even the above-average guys then would flunk. Since Cambridge can't fail so many, they have to increase the number of passes!

Questions set in aptitude and IQ tests on finding the missing digit, number, or letter in a sequence seldom make use of the simple logic or rule required to answer similar questions when set in arithmetic lessons as a time-filler before the bell rings. Here are a dozen of "What's Next" questions that will force you to apply some out-of-the-box thinking.

1 What is the missing digit in this sequence?

$$5, \quad 8, \quad 3, \quad \underline{\quad}, \quad 2, \quad 1$$

2 What is the missing number in this series?

$$0, \quad 1, \quad 4, \quad 15, \quad \underline{\quad}, \quad 325$$

> Don't peep at the hint or answer! Think harder and smarter — your patience and perseverance will pay off!

More Mathematical Quickies & Trickies 15

3 What is the missing number?

90, 96, 43, 56, 39, 50, 68, 89, ___, 78, 94, 55

4 What is the next letter in this sequence?

O, I, Z, E, H, S, G, ___

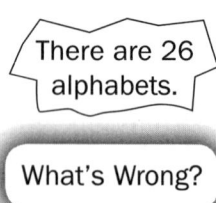

There are 26 alphabets.

What's Wrong?

5 What is the next number in the series?

75, 73, 69, 72, 67, 68, 70, ____

> 1 dollar = 1 cent
>
> *Proof*
> 1 dollar = 100 cents
> = (10 cents)^2
> = (0.1 dollars)^2
> = 0.01 dollars
> = 1 cent

6 What are the next two numbers in this sequence?

1, 4, 3, 11, 15, 13, 17, 24, 23, ___, ___

> Your days are numbered! Why not, "Your days are lettered"?

7 What a the next term in the sequence?

        4   2   8   5   7   1   ?

8 Find the missing number in the following sequence.

        1   4   9   6   5   6   9   9   4   ?

9 What is the missing number in the number series?

$$3 \quad 11 \quad 20 \quad 27 \quad 29 \quad 23 \quad ?$$

10 What is the missing number in the following series?

$$84 \quad 12 \quad 2 \quad \frac{2}{5} \quad \frac{1}{10} \quad ?$$

NCTM vs. Singapore Math

National Council of Teachers of Mathematics (NCTM)

Singapore Math

16 More Mathematical Quickies & Trickies

The Largest Product

Worked Example 1

Using each of the digits 0, 1, 2, 3, and 4 only once, form a 2-digit number and a 3-digit number such that the product is the largest.

Solution

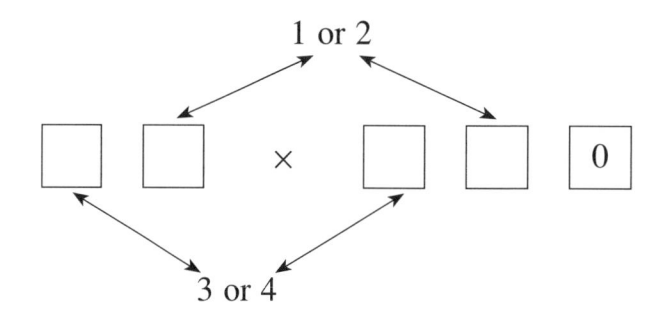

```
         1 or 2
        ↙    ↘
    ↗          ↖
  [ ][ ]  ×  [ ][ ][0]
    ↘          ↙
        ↖    ↗
         3 or 4
```

The first digit in each number can be 3 or 4.
The second digit in each number can be 1 or 2.
Let's take a look at the possible products.

$42 \times 310 = 13{,}020$
$41 \times 320 = 13{,}120$ ←
$32 \times 410 = 13{,}120$ ←
$31 \times 420 = 13{,}020$

Therefore, the largest product is 13,120.

Hence, the two numbers are 32 and 410, or 41 and 320.

> A common mistake is to take the digits in order and expect them to yield the largest product.
>
> $43 \times 210 = 9030$
>
> or
>
> $432 \times 10 = 4320$

Worked Example 2

Use each of the digits 5, 6, 7, 8, and 9 only once, form the largest product.

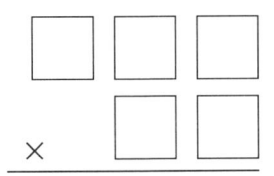

Solution

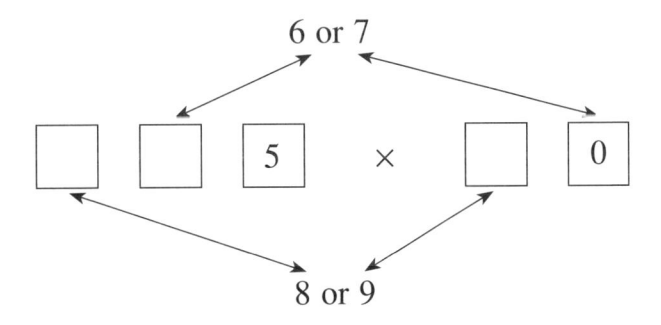

$865 \times 97 = 83,905$
$875 \times 96 = 84,000$ ←
$965 \times 87 = 83,955$
$975 \times 86 = 83,850$

Therefore, the largest product is 84,000.

A common mistake is to simply take

$$\begin{array}{r} 987 \\ \times\ 65 \\ \hline 64,155 \end{array}$$

and assume it to be the largest product.

1 If you took 3 mangoes from a plate holding 13 mangoes, how many mangoes would you have?

2 Insert the signs +, −, ×, ÷ and brackets to make each statement correct.
(a) 1 2 3 4 5 = 10
(b) 1 2 3 4 5 = 50

3 What is the largest 25-digit number that can be divided by 2 and 3 without any remainder?

4 A secretary prints out 6 letters from the computer and addresses 6 envelopes to their intended recipients. In a hurry, her boss stuffs the letters onto the envelopes at random, one letter in each envelope. What is the chance that exactly five letters are in the right envelope?

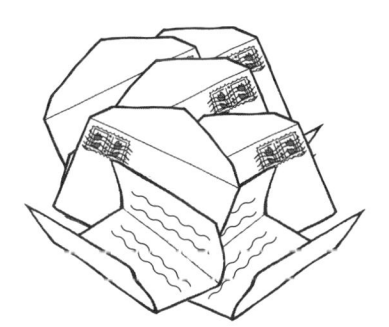

5 If $x + y = 5$ and $x^2 + 3xy + 2y^2 = 40$, what is the value of $2x + 4y$.

6 If 5 dogs dig 5 holes in 5 days, how long does it take 10 dogs to dig 10 holes? Assume that they all dig at the same rate all the time and all holes are of the same size.

7 Solve the following cryptarithm.

$$\begin{array}{r} HAND \\ +NOSE \\ \hline EYES \end{array}$$

8 Show that there will be 8,765,832 hours in this millennium, from the start of 1 January 2000 until the end of 31 December 2999.

9 What is the value of $\dfrac{1}{\sqrt{1}+\sqrt{2}} + \dfrac{1}{\sqrt{2}+\sqrt{3}} + \ldots + \dfrac{1}{\sqrt{24}+\sqrt{25}}$?

10 Four small hexagons are overlapping the large hexagon as shown below. The sides of the large hexagon are twice as long as the sides of the small hexagon. Which area is bigger: The sum of the dotted area of the small hexagons or the shaded area of the non-overlapping part of the large hexagon?

108 More Mathematical Quickies & Trickies 16

17 More Mathematical Quickies & Trickies

What's Wrong — A Comedy of Mathematical Errors

Here is a sample of examples taken from *What's Wrong? – A Comedy of Mathematical Errors*, which contains a collection of errors and mistakes, commonly committed by students, parents, teachers, tutors, writers, and editors.

In each case, can you figure out what is incorrect?

Example 1

If $\sqrt{x}$ has two square roots, -3 and 3, find the value of x.

Solution

Since $3^2 = (-3)^2 = 9$, the value of x is 9.

Example 2

What is x if $x^4 = 16$?

Solution

$x^4 = 16$
Since $16 = 2^4$, so $x = 4$.

Example 3

1 dollar = 100 cents   (1)
$\frac{1}{4}$ dollar = 25 cents   (2)

Taking square roots on both sides:

$$\sqrt{\frac{1}{4}} \text{ dollar} = \sqrt{25} \text{ cents}$$

$$\frac{1}{2} \text{ dollar} = 5 \text{ cents}$$

Where has the missing zero gone to?

Example 4

Express $46\frac{1}{2}\%$ as a decimal.

dec.mal

What's the point of the decimal point?

Solution

$46\frac{1}{2}\% = \frac{46\frac{1}{2}}{100} = \frac{\frac{93}{2}}{100} = \frac{93}{200}$

By long division, $\frac{93}{200} = 0.465$

Example 5

The number 1200 is increased by 25% and then decreased by 25%. What is the new number?

Solution

Since the increase is 25% and the decrease is also 25%, the net effect is zero.

There is no change in the number.

The new number is still 1200.

Example 6

The side of a square is increased by 30%. Find the percentage increase in the area.

Solution

Area = side × side

Percentage increase in area = (0.3 × 0.3) × 100 %
= 0.09 × 100%
= 9%

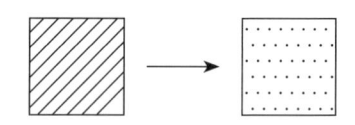

Example 7

A clock takes 6 seconds to strike 4 times. How many seconds does it take to strike 6 times?

Solution

4 times take 6 seconds.

6 times will take $\frac{6 \times 6}{4} = 9$ seconds.

Example 8

A runner takes 9.8 sec to run 100 m. How far will he run in one hour?

Solution

One hour = (60 × 60) seconds

9.8 sec represent 100 meters.

(60 × 60) sec will represent $\frac{100 \times 60 \times 60}{9.8} \approx 36.7$ km.

Example 9

A clock correctly reads 9:30 on a particular Sunday morning, but then starts to run too fast, gaining 4 minutes each hour. What is the actual time when the clock displays 4:30 p.m.?

Solution

From 9:30 a.m. to 4:30 p.m., there are 7 hours.

There is a gain of 4 minutes in one hour.

So, there will be a gain of (7 × 4) = 28 minutes.

Therefore the actual time is 4:30 − 0.28 = 4:02 p.m.

Example 10

How many people must be in a room so that at least 3 of them have the same birthdays?

Solution

The number of people cannot be 3, because they could all be born on different days.

The number of people cannot be 365, because all could be on different days.

So, the number of people must be 3 × 365 = 1095, so that if all the birthdays are spread evenly on every day of the year, we are guaranteed that at least 3 on the same day.

Example 11

Joe walks from home to school at the speed of 4 km/h, and walks back at the speed of 3 km/h. Find his average speed in kilometers per hour for the whole trip.

Solution:

Average speed for the whole trip $= \frac{1}{2} \times (4 + 3)$ km/h
$= 3.5$ km/h

Selected Answers

3. We performed the operations of multiplication and root extraction only on the numbers, and not on the units involved.

7. A wrong answer of 9 seconds results, if we assume that the number of strikes and time are directly proportional to each other.

Since any two strikes are separated by a single time interval, four strikes following one after another are mutually separated by three intervals.

$$\underbrace{\bullet \;\; \overset{2\,s}{} \;\; \bullet \;\; \overset{2\,s}{} \;\; \bullet \;\; \overset{2\,s}{} \;\; \bullet}_{3 \text{ intervals} = 6 \text{ seconds}}$$

Thus, the duration of each interval is 2 seconds.

For 6 strikes, there are altogether 5 intervals between them. Since each interval lasts 2 seconds, 6 strikes will take 10 seconds, and not 9 seconds.

[The actual strokes occupy no appreciable length of time – the 6 seconds are accounted for by the 3 intervals between the 4 strikes. Between 6 strokes there are 5 intervals.]

8. The solution of the runner problem implies that someone, who runs 100 meters in 9.8 seconds, will run over 36 kilometers in an hour.

In practice, a man who runs at such a fast speed is exhausted quickly. Each succeeding 100 meters he runs will be much slower. There is simply no proportionality between the duration of the run at such a high speed and the distance covered.

10. For a non-leap year, there must be $2 \times 365 + 1$ people to be 100% sure of 3 coincidental birthdays or more.
For a leap year, there must be $2 \times 366 + 1$ people.

11. If the distance to be covered is, say, 12 km, and Joseph walks at a speed of 4 km/h, the trip will take 3 hours. At 3 km/h, the same trip will take 4 hours.
It takes 7 hours to cover 24 km, or 1 hour to cover $3\frac{3}{7}$ km.
Average speed for the whole trip $= 3\frac{3}{7}$ km/h

112 More Mathematical Quickies & Trickies 17

1 You are given a strip of paper that is two-thirds of a meter long. Without using a ruler, how long would you cut off a strip half a meter long?

2 What is the next term?

$$3, \ 4, \ 5, \ 5, \ 3, \ 4, \ 4, \ 5, \ 3, \ \underline{\quad}$$

3 What is the minimum difference between two integers that between them contain each digit once?

4 A sheet of paper 0.01 cm is folded fifty times in succession. How thick is the resulting wad of paper?

---

5 Find three different two-digit primes where the average of any two is a prime, and the average of all three is a prime.

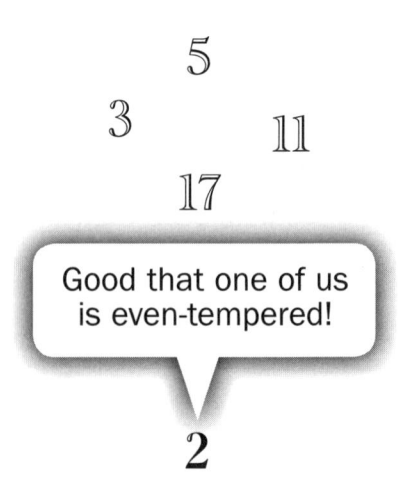

5
3   11
17

Good that one of us is even-tempered!

2

6 If a hen and a half lays an egg and a half in a day and a half, how long (in days) does it take a hen to lay one egg?

7 Find a ten-digit number whose first digit is the number of ones in the number, whose second digit is the number of twos in the number, whose third digit is the number of threes in the number, and so on up to the tenth digit, which is the number of zeros in the number.

8 Using exactly two 2s and any of the standard mathematical symbols, write down an expression whose value is five.

9 The ages of Old and Young total 48. Old is twice as old as Young was when Old was half as old as Young will be when Young is three times as old as Old was when Old was three times as old as Young, How old is Old?

10 A and B are integers that between them contain each of the digits from 0 to 9 once and once only. What is the maximum value of A × B?

18 More Mathematical Quickies & Trickies

The Aha! Myth

Creation occurs in flashes of insight, sometimes called *aha*! reactions.

The late recreational mathematician Martin Gardner suggests ways how one can increase one's chances of solving these "insight" problems, and also of becoming a better problem solver in general.

These problems might seem difficult if one approaches them in traditional ways. But, if one can free one's mind from standard problem solving techniques, one may be receptive to an "aha!" reaction that leads immediately to a solution.

> ***Aha! Creativity***
> - Having no steps
> - Not containing patterns
> - Focusing on big issues
> - Frequently having one's defining moment
> - Using simple methods
> - Being individually intense

How to experience an *aha*! reaction

1. Free yourself from standard, traditional problem solving techniques; forget past experience in order to solve the problems.

2. Sit still, free your mind, and an "aha!" reaction may happen to you.

The "aha!" reaction, if one is lucky enough to receive it, leads immediately to the solution of the problem.

This view of creativity maintains that if one can break away from the hold of past experience one may experience spontaneous solutions to problems.

Insight Problems

1. A man in a certain town in Malaysia married 19 women from the town. All the women are still alive, none has been divorced, the man is neither a Mormon nor a Muslim, and yet the man did not break any laws. How is that possible?

2. **Yesterday, when I went to sleep, I turned off the light and then got into bed. My bed is located six meters from the light switch, and yet I got into bed before the room got dark. How is that possible?**

3. Our basketball team won a game on Friday 13 by the score of 63-52, and yet not one man on our team scored as much as a single point. How is that possible?

4. **What five-letter word do all college graduates spell wrong?**

5. A couple entered a restaurant. They studied the menu, and ordered ten dishes. They were presented with a bill and paid it. When they left, they were as hungry as when they entered! How could that be?

6. **Jill and Roy were hiking in the woods. Jill fell and broke his left leg. Roy wanted to help, but both men knew at once that he could not. Why not?**

7. A man wanted a typical clock with the usual hour and minute hands. However, he wanted its hands to move counterclockwise. Why?

Answers

1. The man is the minister who married the women to their husbands.
2. I went to sleep when it was still daylight.
3. Our team is a women's basketball team.
4. The word is W-R-O-N-G.
5. The couple were planning their wedding reception, which was to be at that restaurant. They selected the food to be served and paid a deposit.
6. Roy was blind, and he would have gone lost if he had left Jill alone.
7. The man was a hairdresser and wanted a backward clock for the rear wall, so that it would appear correct to customers who looked in the mirror.

1 If x is a real number, solve the equation $\sqrt{4-x} = \sqrt{x-6}$.

2 If x and y are positive integers, solve
(a) $(2x + y)(x - 2y) = 7$,
(b) $(x - y)(3x - 2y) = 13$.

Is Wealth × Health a Constant?

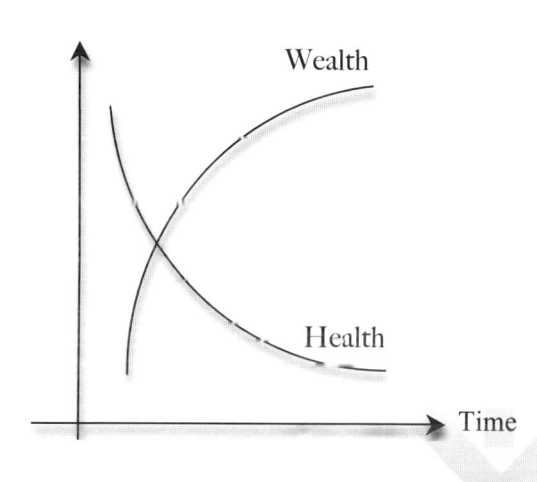

3 Historians believed that Jesus Christ, who was lifted up to heaven on a Sunday, predicted him to return in 25,252,525 days. What day of the week would he be expected to return?

4 Without using a calculator, which of the two numbers is larger?

$$2^{125} \text{ and } 32 \times 10^{36}$$

5 A fussy teenager wears a clean shirt every day. Every Friday morning he drops off the week's laundry and picks up the previous week's laundry. How many shirts must he have?

6 Thirty-six is *interesting* because the sum of its digits (3 + 6 = 9) is equal to its number of letters (not counting the hyphen). How many numbers from 1 to 60, including 36, have a number of letters equal to the sum of their digits? What are these numbers?

7 Look at the network below.

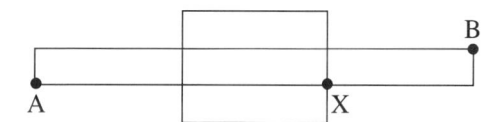

Without passing through any point in the network more than once, how many different paths are there from point A to point B via X?

8 Birthday & Parents' Ages

When Joe was born the year was equal to twice the product of his parents' ages then. When I attended his birthday, the year was equal to the product of his parents' ages then. In what year did I attend Joe's birthday, and how old was he then?

Factorizing recent dates may get you to the answer.

---

9 If $a = \left(1 + \dfrac{1}{2011}\right)^{2011}$ and $b = \left(1 + \dfrac{1}{2011}\right)^{2012}$, find the value of $\dfrac{a^b}{b^a}$.

10 Solve the system of equations.

$$x(x + y) = 9$$
$$y(x + y) = 16$$

Math ART

US$990

US$12,450 *ONLY*

S$1,195

19 More Mathematical Quickies & Trickies

Sam Loyd's *Toughies*

Sam Loyd (1841-1911) was America's undisputed puzzle king, who produced more than 10,000 puzzles in his lifetime. He made a living from creating puzzles, and won many prizes for his puzzles while he was still in school.

In 1858 when he was a teenager he created a puzzle that became an instant classic. His trick donkey's puzzle earned him a cool $10,000 – a huge sum in those days.

Hundreds of thousands of children have played with Loyd's 15-Puzzle*, without realizing who the inventor was. The 1898 puzzle consists of 15 movable blocks which are arranged in a square box in rectangular order, but with the 14 and 15 reversed as shown in the picture.

1	2	3	4
5	6	7	8
9	10	11	12
13	15	14	

The objective is to move the blocks around, one at a time, to bring back to the original position so that 14 comes before 15, or that 13, 14 and 15 are in order. The puzzle created such a craze in America and Europe that employers had to ban it during office hours.

> Is this problem solvable?

> $1000 for the correct solution!

You can be a modern-day Sam Loyd, by inventing creative puzzles and brain-twisters. Here are five *toughies* designed by America's puzzle king.

*It is believed that Loyd lied that he invented the 15-puzzle. See Jerry Slocum's *The 15 Puzzle: How it Prove the World Crazy*.

1. When the hour and minute hands are at equal distance from the six hour, what time will it be exactly?

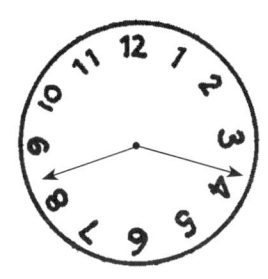

2. How much does the baby weigh if the mother weighs 100 pounds more than the combined weight of the baby and dog and the dog weighs 60 percent less than the baby?

3. Arrange these seven figures and the eight dots so they would add up to 82.

    • 4 • 5 • 6 • 7 • 8 • 9 • 0 •

4. Arrange 9 matches to make ten, and 6 matches to make nothing.

5. 5219 votes were cast at a recent election for the four candidates running for a particular office. The winner received 22 more votes than candidate A, 30 more than candidate B, and 73 more votes than candidate C. How many votes did the winner receive?

Answers

1. Exactly 8 h $18\frac{6}{13}$ min or 8 h 18 min $27\frac{13}{9}$ s.

2. Mother: 135 pounds, baby: 25 pounds, dog: 10 pounds.

3. A dot over a decimal fulfills the same purpose as a straight line, making the decimal infinitely repeat.

   80
   $.\dot{5}$
   $.\dot{9}\dot{7}$
   $+.\dot{4}\dot{6}$
   $\overline{82}$

4. Nine matches form the word TEN, and six to form the word NIX.

5. If x represents the winner's number of votes, then
   $x + A + B + C = 5219$
   $x + (x - 22) + (x - 30) + (x - 73) = 5219$
   $4x = 5344$
   $x = 1336$

1 Solve the equation $\sqrt{x+5} - \sqrt{x} = 1$.

2 From a chain of 23 paper clips, if I remove two of them, I can make a chain of any length from one to 23 out of the five pieces. Which two should I remove?

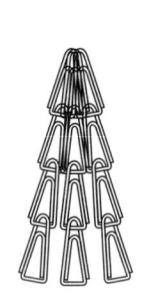

3 Solve the equation $\sqrt{x+2} = -2$.

4 How can you obtain the number 3 using only zeros and common mathematical symbols?

5 Two positive integers a and b are such that the difference of $a \times b$ and $a + b$ is 1000. If a is a perfect square, what is the value of b?

6 Solve $\sqrt{2x+3} + \sqrt{x+3} = 0$.

7 What is the next term in the sequence?

$$1, \quad 2, \quad 6, \quad 12, \quad 60, \quad 60, \quad 420, \quad 840, \quad ...$$

8 Solve $\sqrt{\dfrac{3-x}{2+x}} + 3 \times \sqrt{\dfrac{3-x}{2+x}} = 4$.

9 At a meeting there are 100 politicians. Each one is either a liar or truth-teller. At least one is honest. Given any two of the politicians, at least one is dishonest. How many are honest and how many are dishonest?

10 If x is an integer, solve the equation

$$x^5 - 101x^3 - 999x^2 + 100{,}900 = 0.$$

20 More Mathematical Quickies & Trickies

The Tuesday Boy Problem*

by Yan Kow Cheong on Thursday, July 15, 2010 at 11:39am

A blogworthy topic around the world, at least among recreational mathematicians, is the Tuesday boy problem:

I have two children. One is a boy born on a Tuesday. What is the probability I have two boys?

What has the day of the week got to do with the question? Everything, according to the problem poser of this *mathematical quickie* or *toughie*.

The key to solving the problem is to properly interpret the question. What Gary Foshee, a collector and designer of puzzles, meant was: Of all the families with one boy and exactly one other child, what proportion of those families have two boys?

Give it a try before you read on. What's your answer? Did you get 1/4? Or 1/3?

Let's consider answering a similar but simpler question so that we can pin down what Foshee's question is really asking.

I have two children. One of them is a boy. What is the probability I have two boys?

For two children, there are four equally likely possibilities: BG, GB, BB or GG.

One condition for this question is that one child must be a boy. So we can exclude the GG case, leaving us with only three options: BG, GB and BB. Since one out of these three cases is BB, the probability of the two boys is 1/3.

---

*Originally posted on Facebook.

Now let's use the same reasoning for the original question, by listing the equally likely possibilities of children, together with the days of the week they were born in. Let's call a boy born on a Tuesday a BTu. There are four possible cases:

1. When the first child is a BTu and the second is a girl born on any day of the week: there're 7 different possibilities.

2. When the first child is a girl born on any day of the week and the second is a BTu: again, there're 7 different possibilities.

3. When the first child is a BTu and the second is a boy born on any day of the week: again there're 7 different possibilities.

4. When the first child is a boy born on any day of the week and the second child is a BTu: there are 7 different possibilities here too, but one of them —when both boys are born on a Tuesday—has already been counted when we considered the first to be a BTu and the second on any day of the week. So, since we are counting equally likely possibilities, we can only find an extra 6 possibilities here.

There are a total of 7 + 7 + 7 + 6 = 27 different likely combinations of children with different genders and birth days, of which 13 of these are two boys. So the answer is 13/27, which is different from 1/3.

Probabilistically yours

KC - *Keep Counting*

1 A couple plans to shop at 3 shopping malls in Singapore, followed by 4 selected malls in Hong Kong, followed by 5 selected shopping centers in Dubai. In how ways can the couple order their itinerary?

2 A licence plate consists of three digits (0 through 9), followed by a letter (*A* through *Z*), followed by another digit. How many different licence plates are possible?

512X6

3 The inhabitants of Topoland are type A or type B. Type A people can ask only questions whose correct answer is "yes." Those of type B can ask only questions whose correct answer is "no." Rick was overheard to ask, "Are Ruth and I both type B?" Which type is Ruth?

4 How many subsets does a 100-element set have?

5 Find the smallest integer n such that the number of subsets of an n-element set is greater than 10^{100} (a googol).

6 A monk started up a hill one morning and reached the top at sunset. He stayed overnight in the temple there, started back down the hill the next morning by the same path, and arrived at the bottom at sunset. Was there a place on the path that he passed at exactly the same time of day going both ways?

7 How many strings of 10 symbols are there in which the symbols may be 0, 1, or 2?

8 Lucy wants to arrange 4 science books, 5 fiction books, and 6 history books on a shelf. Books of the same category are to be grouped together, but otherwise the books may be put in any order. How many orderings are possible?

9 Which numbers (0, 1, ..., 9) each of the letters stands for?

$$\begin{array}{r} HALF \\ +HALF \\ \hline WHOLE \end{array}$$

> There is more than one solution.

10 A fisherman leaves his boathouse on the river and rows upstream at a steady rate. After 2 km he passes a log floating down the river. He continues on for another hour, and then turns around and rows back downstream. He overtakes the log just as he reaches the boathouse. What is the flow speed of the river?

21 More Mathematical Quickies & Trickies

What Is 1 + 1, Really?

Mathematically speaking	$1 + 1 = 2$
Synergistically speaking	$1 + 1 > 2$
Prayerfully speaking	$1 + 1 \neq 2$
Philosophically speaking	$1 + 1 = 3$ for large values of 1
Politically speaking	$1 + 1 = 110$
Unscrupulously speaking	$1 + 1 = 1.95$
Conjugally speaking	$1 + 1 = 1$
Statistically speaking	$1.95 < 1 + 1 < 2.05$
Paradoxically speaking	$1 + 1 = 4$
Analytically speaking	$1 + 1 \to 2$
Probabilistically speaking	$1 + 1 = \dfrac{1}{\frac{1}{1+1}}$
Approximately speaking	$1 + 1 \approx 2$
Chemically speaking	$1 + 1 = 2 - \text{heat}$
Pessimistically speaking	$1 + 1 < 1$
Limitly speaking	$1 + 1 = 1.9999\ldots$
Unfairly speaking	$1 + 1 > 1 \times 1$
Productively speaking	$1 + 1 = 300\%$
Lovingly speaking	$1 + 1 = \text{Anything}$
Blindly speaking	$1 + 1 = \cdot\,\cdot$
Alphametically speaking	ONE + ONE = TWO
Deceptively speaking	$1 + 1 > 1.999\ldots$

Magically speaking	$1 + 1 = 0$
Boastfully speaking	**$1 + 1 = 2.000\ldots$**
Promotionally speaking	$1 + 1 = 2 + 1$ free
Truthfully speaking	**$1 + 1 < 2$**
Joyfully speaking	$1 + 1 \geq 3$
Exploitatively speaking	**$1 + 1 = \frac{1}{2}$**
Rigorously speaking	$1 + 1 = 2$ (after 200 pages of proof)
Internally speaking	**$1 + 1 = 10$**
Averagely speaking	$1 + 1 = 1$
Blessedly speaking	**$2 < 1 + 1 < \infty$**
Gracefully speaking	$1 + 1 > 2$
Mercifully speaking	**$1 + 1 \leq 0$**
Topologically speaking	$1 + 1 = 4$
Lawfully speaking	**$1 + 1 = 1.8$**
Numerally speaking	$1 + 1 = $ II
Erroneously speaking	**$1 + 1 = \sqrt{1^2 + 1^2}$**
Axiomatically speaking	$1 + 1 = ?$ *Refer to axioms 1 – 14*
Passionately speaking	**$1 + 1 = 5$**
Artistically speaking	$1 + 1 = 1.7$
Mergingly speaking	**$1 + 1 = 1 +$ layoff**
Computingly speaking	$1.0000\ldots + 1.0000\ldots = 1.000\ldots$ e
Hintfully speaking	**$1 + 1 = ?$ See Appendix.**
Murphologically speaking	$1 + 1 $ ☞$ 2$, where ☞ symbolizes for "hardly ever."
Creatively speaking	**$1 + 1 = 11$**
Biologically speaking	$1 + 1 = 3$

1 Fill in the missing numbers.

2	5		17		37
50	65		101		

2 In the equilateral triangle two matchsticks meet at each vertex. Use 12 matchsticks to form an arrangement in which three matchsticks meet at each vertex.

3 Dave and Esther are 70 km apart along a straight line. They start cycling towards each other's starting point at the same time. Dave's speed is 16 km/h and Esther's speed is 24 km/h. How long later will they be 10 km apart?

4 Find two-digit positive integers in such that n is twice of its digital product.

5 A cereal company places toys in its cereal boxes. There are four different toys distributed evenly over all the boxes the company produces. On average, how many boxes of cereal would you need to buy before you collected a complete set?

6 Two rods each 20 cm long are joined at one end until they are 24 cm apart. The rods are spread until they are 32 cm apart.
(a) What is the area of the triangle in each case?
(b) As the rods are spread, the area of the triangle will get bigger to a point, then fall. At some point the area will be maximum. What will be that maximum area?

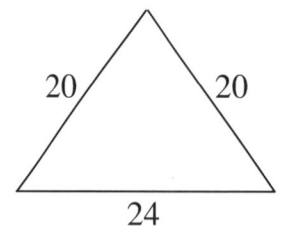

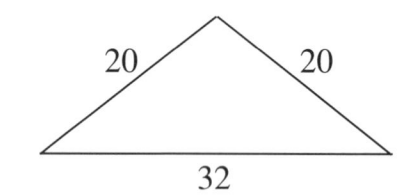

7 When the clock shows 3:36 the product of the hours and minutes is 3 × 36 = 108, which is the number of degrees between the two hands. When would be the next two times when this happens again?

8 Each of the railroad stations in a given area sells tickets to every other station on the line. This practice was continued when several new stations were added, and 52 additional sets of tickets had to be printed. How many stations were there initially, and how many new ones had to be added?

9 Using all the digits from 1 to 9 only once, form three 3-digit numbers that are in a ratio of 1 : 2 : 3.

For example, 192 : 384 : 576 = 1 : 2 : 3

There are 4 solutions.

10 The number 121 is a palindrome and is also the square of a palindrome (11) and the square root of a palindrome (14,641). What is the next number of this type?

Madam I'm Adam.

Palindromically, yours, Sir!

22 More Mathematical Quickies & Trickies

In Love with Cryptarithms

A **cryptarithm** is a puzzle in which the digits are replaced by letters. The challenge is to find the missing digits representing the letters. Using some properties of numbers and applying logical reasoning we can solve these puzzles in a more structured manner.

Even laypersons would benefit from toying with cryptarithms, as these number-word puzzles will not only help them to improve their logical thinking skills but also enhance their chances of tackling them effectively, as they often appear in aptitude tests and mathematics contests.

Let's apply some strategies in solving a few cryptarithms intelligently.

Example 1

In the addition on the right, since A and H represent two different letters, we need to find the values of A and H.

```
   A
   A
 + A
 ---
  HA
```

In the ones or units column, A + A + A = A. So A can be either 0 or 5.

Note that the sum A + A + A is also a two-digit number. This excludes the case for A = 0, as 0 + 0 + 0 = 0.

```
   5
   5
 + 5
 ---
  15
```

The sum HA could have been 00, but that would not fulfill the condition that H and A are different letters. Therefore, A must be 5.

Since A = 5, we can replace the value into the cryptarithm to find the value of H.

Example 2

Let's look at a cryptarithm involving multiplication:

$$\begin{array}{r} AB \\ \times\ C \\ \hline AAA \end{array}$$

If A = 1, then
$$\begin{array}{r} 1B \\ \times\ C \\ \hline 111 \end{array}$$

Now, 111 is not the multiple of any two-digit number, except 37. That would make C = 3 and A = 3, but A and C stand for two different digits.

If A = 2, then
$$\begin{array}{r} 2B \\ \times\ C \\ \hline 222 \end{array}$$
but 222 is not the multiple of any two-digit number.

If A = 3, then
$$\begin{array}{r} 3B \\ \times\ C \\ \hline 333 \end{array}$$
. Since 37 × 9 = 333, so one answer is
$$\begin{array}{r} 37 \\ \times\ 9 \\ \hline 333 \end{array}$$

The same reasoning is applied to A = 4, 5, …, 9. No corresponding values of B and C satisfy the given pattern. Hence the only solution is 37 × 9 = 333.

$$\begin{array}{r} AB \\ \times\ C \\ \hline AAA \end{array} \longrightarrow \begin{array}{r} 37 \\ \times\ 9 \\ \hline 333 \end{array}$$

Example 3

Solve the alphametic: $(HE)^2 = SHE$.

Since both HE and SHE end in the same digit, E can take one the following: 0, 1, 5 or 6.

Since SHE is less than 1000, HE cannot exceed 32 ($\sqrt{1000} \approx 31.6$). That leaves 9 possibilities: 10, 15, 16, 20, 21, 25, 26, 30, and 31.

The only possibility turns out to be HE = 25, so that $25^2 = 625 =$ SHE.

Hence the solution to $(HE)^2 =$ SHE is $25^2 = 625$.

Practice

1. In each question, the letters represent digits. No digits are repeated in each question. Find the value of each letter.

   (a) A   (b) P   (c) XY   (d) KLK   (e) BA
      + B   Q   + Y   − MK   × 7
      ———   + R   ———   ———   ———
       CC   ———   YX   KL   HAA
               PR

2. Solve the following: ABCDE
                          ×   4
                          ———————
                          EDCBA

3. Solve the alphametic: AREA = PI × R^2

4. Solve the following cryptarithms.

   (a)   SEND   (b)   SPEND
       + MORE   − MORE
       ——————   ——————
        MONEY   MONEY

Selected Answers

(a) Since CC must be equal to 11 (as A + B must be less than 22), the possible solutions are:

9	8	7	6	5	4	3	2
+2	+3	+4	+5	+6	+7	+8	+9
=11	=11	=11	=11	=11	=11	=11	=11

(b) Since the units digit of P + Q + R is R, the units digit of P + Q is 0, so P + Q = 10.
Since P − 1, Q − 8, R is any other one digit number 7, 3, 4, 5, 6, 7 or 8

(c) X + 1 = Y; X is even, say, 12, 14, 16 or 18, then Y = 6, 7, 8 or 9

   X = 8   89
   Y = 9   + 9
            ———
             98

(d) K = 1   L = 0   M − 9 = 1
               16 −   10
               91
               ———
               101

(e) When A is multiplied by 7, what results is a number whose ones digit is A.
This implies that A = 0 or 5.
If A = 0, then B = 0. But A and B are different digits. So A is rejected; A must be 5.   BA
Since B times 7 plus a 'carry' of 3 also ends in 5, B is 6. Then the multiplication BA × 7 = HAA   × 7
becomes 65 × 7 = 455.   ———
                                                                                                   HAA

2. 21,978 × 4 = 87,912

3. 4704 = 96 × 7^2.

1. A clock takes 3 seconds to strike 3 o'clock. How long does it take to strike 6 o'clock?

2. At a dinner, there are 3 choices for the appetizer, 4 choices for the main course, and 2 choices for the desert. How many different meals are possible?

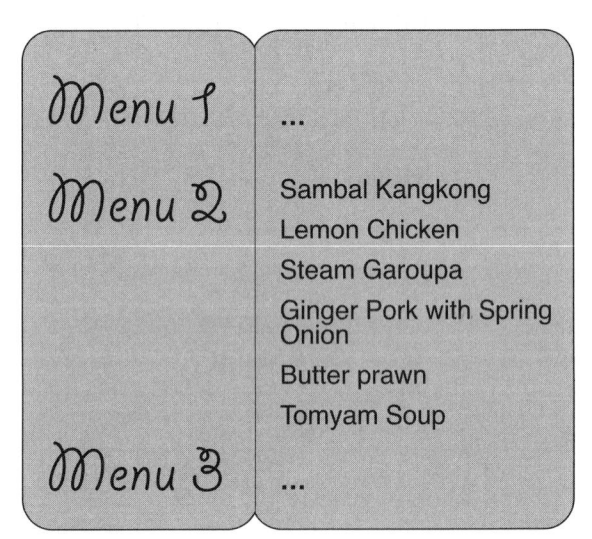

Menu 1 ...
Menu 2
- Sambal Kangkong
- Lemon Chicken
- Steam Garoupa
- Ginger Pork with Spring Onion
- Butter prawn
- Tomyam Soup

Menu 3 ...

3. A given weight can balance either 5 silver coins or 6 gold coins. One side of the scale pan containing 5 such weights can balance 12 gold coins and some silver coins. How many silver coins are there?

4. Christmas Day, 1902 was a Thursday.
   (a) Which day of the week was Christmas Day, 1903?
   (b) Which day of the week was Christmas Day, 1904?

5 Ian has 5 small balls, 4 medium-sized balls, and 3 large balls. If balls of the same size cannot be distinguished, how many ways can he arrange the balls in a row?

6 Find the smallest integer n such that $1 \times 2 \times 3 \times \cdots \times (n-1) \times n > 10^{100}$ (a googol).

7 Solve the alphametic:

$$A \overline{)\text{MERRY}}^{\text{XMAS}}$$

148 More Mathematical Quickies & Trickies 22

8 How many subsets of the set $\{a, b, c, ..., x, y, z\}$ do not contain the letters c, h, r, i, s, t?

9 A packet of candies was on sale for 96 cents. The percentage profit made by the shopkeeper had the same numerical value as the cost, in cents, to him. How much did it cost the shopkeeper?

10 Determine the sum of the following:

$$1 + \frac{1}{3} + \frac{1}{6} + \frac{1}{10} + \frac{1}{15} + \cdots$$

Can I use a visual (look-see) proof to find the sum?

Thanks to Global Warning

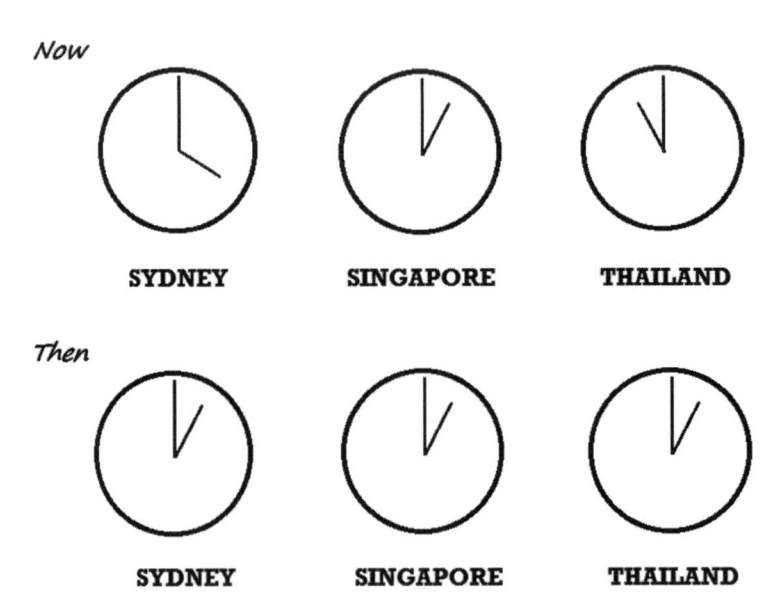

Now

SYDNEY SINGAPORE THAILAND

Then

SYDNEY SINGAPORE THAILAND

23 More Mathematical Quickies & Trickies

Mathematical Kiasuism*

The *mathematical kiasuist* (usually a male) has some of these noticeable characteristics:

1. **He hides books so that others cannot access them.**
2. He goes all the way out to borrow the seniors' notes so that he can read them before the academic term starts.
3. **He does not have a girlfriend, nor does he plan to have one.**
4. He is conspicuously seated on the first row of the lecture hall.
5. **He keeps on renewing the library books until he can no longer do so (even though he might not need them.)**
6. He photostats the past exam papers before the next academic term starts.
7. **He is reading his first borrowed university lecture notes while he is serving his national service.**
8. He usually wears thick glasses and carries a few (thick) books each time.
9. **He is seen to be reading journals and dissertations, although he might not understand their content.**
10. He is usually the few odd ones to approach the lecturers for additional materials to lay his hands on.
11. **He appears to be helpful when approached although he heartedly prefers not to be disturbed.**
12. He is generally up-to-date with current issues not related to his field.
13. **He seldom gives private tuition, as he would rather spend the time updating and upgrading himself.**

*Traits of *kiasuism*: Afraid to lose out, displaying signs of self-centeredness and selfishness.

1 A car driver travels a certain distance at 60 km/h and arrives at his destination one hour earlier than if he had driven at 50 km/h. What was the distance?

2 It takes 5 men 6 hours to dig 7 holes. How long does it take one man to dig half a hole?

3 The sum of the squares of two consecutive numbers is 1105. What are the two numbers?

4 The square of 45 is 2025 and 20 + 25 = 45, the number we started with. Find two other numbers with 4-digit squares that exhibit the same peculiarity.

5 A book has n consecutive pages (each page has 2 sides, each side is one page) torn out. If a is the last numbered page before the torn-out section and b is the first numbered page following the torn-out section, express n in terms of a and b.

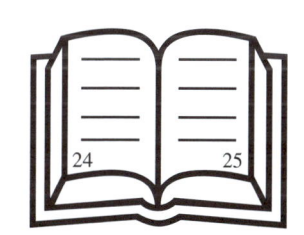

Looks like that vandal tore away the *Answers* pages!

6 What is the least positive integer such that
(a) dividing by 7 giving a remainder of 4,
(b) dividing by 8 giving a remainder of 5,
(c) dividing by 9 giving a remainder of 6?

I ~~don't~~ like math.
I ~~don't~~ like math.
I ~~don't~~ like math.

7 Use eight distinct digits among 0, 1, 2, ..., 9 to form the product

☐☐☐
× ☐
─────
☐☐☐☐

154 More Mathematical Quickies & Trickies 23

8 Insert the appropriate brackets to make the following statement true.

$$5 - 2 \times 1 + 4 \div 6 = 5$$

---

9 Find the largest group of different positive integers less than 100 such that no combination of them added together totals 100.

---

10 Two integer solutions of the equation $a^2 + b^2 = c^2$ are 3, 4, 5; and 5, 12, 13. Find a three-dimensional example of a rectangular parallelepiped whose edges and diagonal can be expressed as integers, that is, an integral solution of $a^2 + b^2 + c^2 = d^2$.

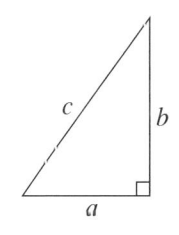

(a, b, c) is a Pythagorean triple.

More Mathematical Quickies & Trickies

The Mathemagic of 142857

(a) Express 999,999 as a product of prime factors.

(b) Express 142,857 as a product of prime factors.

(c) Express the fraction $\frac{142,857}{999,999}$ in lowest terms.

(d) Write down the first 7 multiples of 142,857.

(e) What are the other sevenths expressed as fractions with a denominator of 999,999?

(f) Write down each of the sevenths as recurring decimals.

Solution

142,857 is a "magical" number since it appears in the expansion of the $\frac{1}{7}$'s as recurring decimals.

(a) $999,999 = 3^3 \times 7 \times 11 \times 13 \times 37$

(b) $142,857 = 3^3 \times 11 \times 13 \times 37$

(c) $\frac{142,857}{999,999} = \frac{1}{7}$

(d) The first 7 multiples of 142,857 are:

$1 \times 142,857 = 142,857$
$2 \times 142,857 = 285,714$
$3 \times 142,857 = 428,571$
$4 \times 142,857 = 571,428$
$5 \times 142,857 = 714,285$
$6 \times 142,857 = 857,142$
$7 \times 142,857 = 999,999$

(e) $\frac{285,714}{999,999} = \frac{2}{7} = .285714285714\ldots$

$\frac{428,571}{999,999} = \frac{3}{7} = .428571428571\ldots$

$\frac{571,428}{999,999} = \frac{5}{7} = .571428571428\ldots$

$\frac{714,285}{999,999} = \frac{5}{7} = .714285714285\ldots$

$\frac{857,142}{999,999} = \frac{6}{7} = .857142857142\ldots$

1 It is Sunday. John the Baptist goes out into the wilderness, due to return after forty days. What day of the week will he return?

2 What is the missing number?

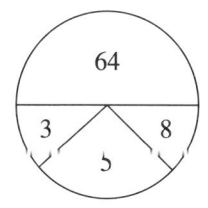

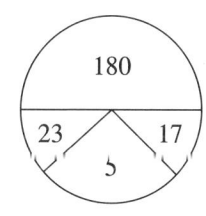

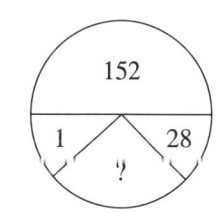

3 Write 55 by using five 4's.

More Mathematical Quickies & Trickies 24

4 What are two smallest whole numbers where the difference of their squares is a cube, and the difference of their cubes is a square?

5 What is the largest three-digit prime each of whose digits is a prime?

6 If Jeremy's age is multiplied by his father's age, the product is a permutation of the digits in their individual ages. How old are they?

7 Diophantus's boyhood lasted $\frac{1}{6}$ of his life. He grew a beard after $\frac{1}{12}$ more. After a $\frac{1}{7}$ more of his life he is married. Five years later he had a son. The son lived $\frac{1}{2}$ as long as his father, and the father died 4 years after his son. How old was Diophantus when he died?

8 Use all ten digits (0, 1, 2, ..., 9) to form a mixed number, which can be simplified to an integer.

9. What is the largest *n*-digit number which is also an exact *n*-th power?

10. What is the smallest number that, when divided by successively by 45, 454, 4545, and 45,454 leaves the remainders 4, 45, 454, and 4545 respectively?

25 More Mathematical Quickies & Trickies

The Lighter Side of Singapore Math

Our Singapore Heritage

The Greeks have their Euclid's *Elements*.
The Egyptians had their Rhind papyrus.
The Japanese had their *sangaku*.
The Chinese had their *Chiu-chang Suan-shu*.

What do the Singaporeans have?
The Model Method

Evolution of Singapore Grade 5/6 Math Word Problems

Teaching Math in 1970 (Traditional Math)

Mr. Yan has three cows and eight chickens. How many legs are there all together? [Answer from parent]

Teaching Math in 1980 (Model Method)

Mr. Yan has twice as many cows as chickens. If there are 64 legs in all, how many cows are there? [Answer from teacher]

Teaching Math in 2000 (Problem Solving)

Mr. Yan has cows and chickens. If there are altogether 28 legs, how many chickens and how many cows are there? [Answer from tutor]

Teaching Math in 2010 (Challenging)

Mr. Yan has almost twice as many chickens as cows. The total number of legs and heads is 184. How many cows are there? [Answer from trainer, paid website, or *Ask Dr. Math* website]

1 Fill in the boxes.

1	2	4		10	12		18
	28	30		40	42		

---

2 If $0 < b < a$ and $a^2 + b^2 = 6ab$, find the value of $\frac{a+b}{a-b}$.

---

3 Find $p, q \in \mathbb{R}$ such that $p + q = pq$.

4 Find two numbers whose difference, sum, and product are in the ratio of 1 : 4 : 15.

Christmas Day in 2025 is on a Thursday!

I'd prove Santa exists statistically!

5 Make the number 500 by mathematically combining eight 4's.

6 Find distinct digits A, D, E, F, G, I, N, S, T from {0, 1, 2, ...} with A, G, S ≠ 0 such that

$$\text{SEND} \div \text{A} = \text{GIFT}.$$

7 If $(a + b) : (b + c) : (c + a) = 6 : 7 : 8$ and $a + b + c = 14$, find the values of a, b, and c.

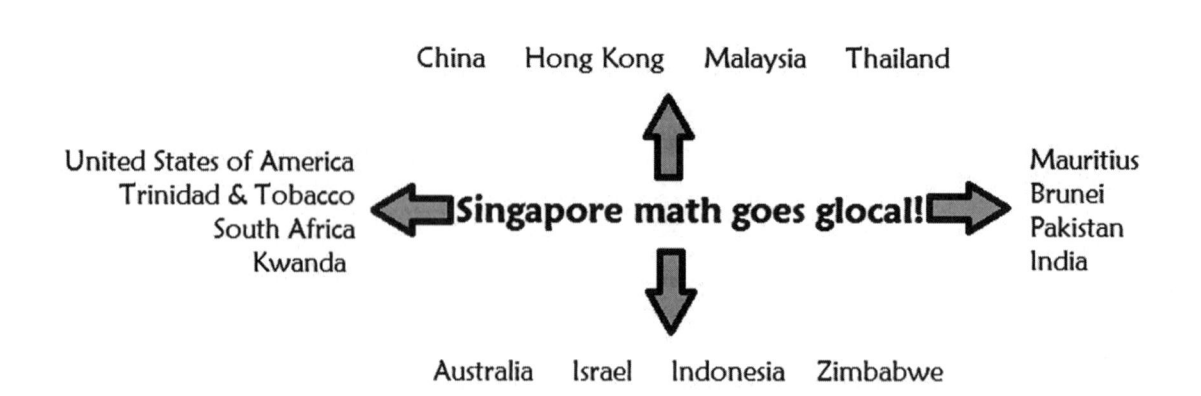

8 Use all ten digits (0, 1, 2, ..., 9) to form a mixed number equal to 299.

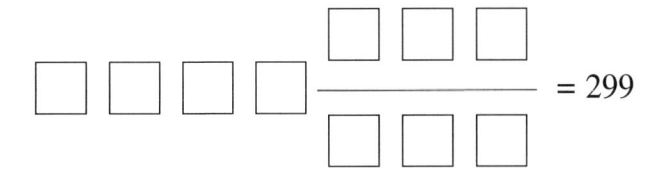

$$\square\square\square\square \frac{\square\square\square}{\square\square\square} = 299$$

9 Find distinct digits *a*, *b*, *c*, *d*, *e* from {1, 2, …, 9} such that the multiplication

$$\begin{array}{r} a\ b\ c\ d\ e \\ \times\qquad\quad 4 \\ \hline e\ d\ c\ b\ a \end{array}$$

10 Triangle *ABC* is a right triangle with integer side-lengths whose area is twice of its perimeter. What are the dimensions of the sides?

The Google Mythology

Beware, I'm super-addictive!

www.google.com

And I'm more than just a search engine.

Google is a play on the word **googol**, the number 1 followed by 100 zeros.

I use Google as my calculator, by typing a sum into the search box and Google will give me the answer.

Google's my dictionary now! I just type "define" before the word I want to now.

26 More Mathematical Quickies & Trickies

K C Yan's Laws & Lores

If you doubt your answer may be wrong, it is probably wrong.

If you are taking less time to solve an optional question, you have probably used the wrong method.

If you have no time to complete your math paper, you are probably below average, at best.

Leave any creative solutions for non-examination settings.

You will take as much time (or more) to answer the last 20 percent part of a question than the first 80 percent of the question.

To score A is easy, but to score A* is tedious.

Sounds like King Solomon talking!

There is nothing mathematically new under the sun.

All 'new' mathematics are rehashed old mathematical news.

A creative solution is the aftermath of hundreds of hours of frustration, bucketfuls of perspiration, and a few minutes of exaltation.

Mathematical problem solving is an art, not a science.

Most in-service problem solving courses for mathematics teachers are a waste of time and money.

I prefer problem posing to problem solving.

There is no short cut to mathematical problem solving.

What is likely to go mathematically wrong will go wrong.

A mathematician is a juggler of notions and notations – ideas and symbols.

1 In the following, what is the hidden phrase or title?

```
        4      12
   17
              62
        113
                56
         GOOD
```

---

2 If $\left(\dfrac{a+b}{2}\right)^2 - ab = 25$, what is the value of $\dfrac{a^2+b^2}{2} - \left(\dfrac{a+b}{2}\right)^2$?

3 Solve $x(x + 4) + \frac{1}{x}\left(\frac{1}{x} + 4\right) = 10$.

4 Find all two-digit positive integers in such that their values are increased 75% when the digits are reversed.

5 The digit 3 is written at the end of a two-digit integer such that the difference between the new integer and the original integer is 372. Find the original integer.

6 Find the smallest positive integer n such that $n \times 999$ does not contain the digit 9.

7 If $x^2 = xy - 1$ and $y^2 = 1 - y$, show that $x \neq 1$ and $x^5 = 1$.

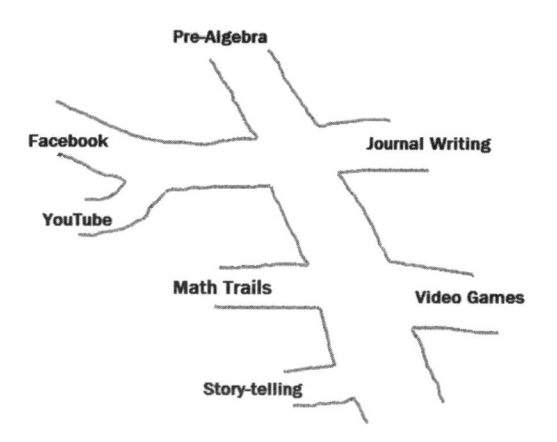

8 If $x + \dfrac{1}{x} = 2$, find the values of $x^2 + \dfrac{1}{x^2}$, $x^3 - \dfrac{1}{x^3}$, $x^4 + \dfrac{1}{x^4}$, $x^5 - \dfrac{1}{x^5}$, and $x^{2n} + \dfrac{1}{x^{2n}}$, $x^{2n-1} - \dfrac{1}{x^{2n-1}}$ for any positive integer n.

9 Find distinct digits A, B, C, H, I, J, K from {0, 1, 2, …, 9} with A, B, C ≠ 0 such that

$$\begin{array}{r} A\,A\,A \\ B\,B\,B \\ +\ C\,C\,C \\ \hline H\,I\,J\,K \end{array}$$

10 Find $x, y \in \{0, 1, 2, …, 9\}$ in $2xy89 = n^2$ for some $n \in \mathbb{Z}^+$.

27 More Mathematical Quickies & Trickies

Flee and Free from the FREE

It is almost an irony that Singapore, which is highly regarded for its mathematical performance among both developed and developing countries, has a high rate of innumeracy among its citizens. *Innumeracy* is the mathematical equivalent of illiteracy—when someone refuses, or is unable, to think mathematically.

Our schools continue to produce a large number of drill-and-practice (or drill-and-kill) specialists every year. Most are brought up on model answers and exam techniques, rather than on being trained to think creatively and critically. As a result, many citizens are more easily prone to numerical terrorists—charlatans who prey on their innumeracy.

Just like there are on-going health talks being held during the weekends to educate the public in adopting a healthier lifestyle, there could also be quantitative talks or seminars to warn the public against the abuses by numerical terrorists.

Nowhere is innumeracy more rampant than in the retail industry. Below are some advertisements that are targeted at those who think they are getting a good deal.

Maximum:
3 per customer
Only $9.95

Buy TWO

and get ONE free.

LAST DAY
Usual Price ~~$135~~
NOW $27.90

---

3 for only $10
1 for $4.50

SUPER BUYS!

SAVE 25% TO 90%

3 DAYS ONLY

First 200 registrations get a printer for only $1.00!

It is not surprising that many people succumb to many high-tech frauds, many of which promoting supposedly free-of-charge goods and services. Here are some common numerical deceits.

- **Subscribing to a magazine by a certain date to be entitled to discounts and free gifts**
- Collecting some free air tickets after watching some video ads
- **Banking (0% interest rate for easy credit lines and mutual funds, …)**
- Attending a free talk for tips on becoming rich
- **Collecting "free gifts" for the first 100 customers**
- Meeting agents promising no-obligation to buy goods or services
- **Inspecting your house for termites for free**
- Fumigating your entire house or flat for free
- **Instaling some high-tech security alarm to guard against burglary and terrorism**
- Registering for some mathematics seminars by some world-renowned professors (Vedic Math, Speed Math,…)
- **Upgrading your insurance premium against elephantiasis, SARS, or some rare diseases, at no extra cost**
- Insuring yourself and your family against comets, earthquake, terrorism and biological warfare at no cost
- **Telecommunications ("free" IDD phone rates, "free" talk time, …)**
- Timesharing (staying in castles and fortresses, and privately-owned islands at no extra cost)
- **Retailing: enjoy-first-and-pay-later schemes, "free" tours, "free" vouchers, …**
- Gambling (cheaper on-line bets, "free" *Luohan* fish for divining winning lottery numbers, …)
- **Taking part in pyramid selling**
- Investing in some get-rich schemes from VIPs (Nigerian, Chinese, Russian conmen, …)
- **Applying for "interest-free" credit cards**

The last thing these allegedly free offers or services would do to the gullible would be to set them free from any financial and psychological freedom. So, let us stay FREE-free.

1. Fill in the boxes.

(a)
1	2	4		10	12		18
	28	30		40	42		

(b)
12	5		17	37
30	65		101	

2. Write 55 by using five 4's. Make the number 500 by mathematically combining eight 4's.

3. Use all ten digits to form a mixed number which is equal to 299.

$$\square\square\square \, \frac{\square\square\square\square}{\square\square\square\square} = 299$$

4 Find two 2-digit numbers with a sum of 100 and a product of 2451.

5 For any positive integer n, find all positive integers p and q which such that

$$\frac{1}{p} + \frac{1}{q} = \frac{1}{n}.$$

6. A six-pointed regular star is made up of two interlocking equilateral triangles. What is the ratio of the area of the entire star to the area of one of the equilateral triangles?

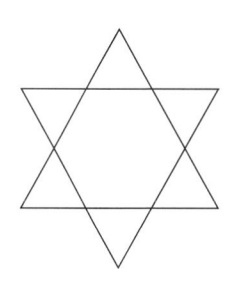

7. If $\left(x + \dfrac{1}{x}\right)^2 = 3$, find the value of $x^3 + \dfrac{1}{x^3}$.

8. Solve the system of simultaneous equations

$$x + y + z = w, \frac{1}{x} + \frac{1}{y} + \frac{1}{z} = \frac{1}{w}$$

9. The area of a rectangle is 3 times of its perimeter and side-lengths are integers. What is its area?

10 Given 12 toothpicks, how many ways can you arrange any number of toothpicks, end-to-end, to form a triangle?

For example, a 3-3-4 arrangement can form an isosceles triangle, with two sides 3 toothpicks long and a base 4 long. Note that 3-4-3 or 4-3-3 give the same triangle as 3-3-4.

Valentine's Day For Nerds!

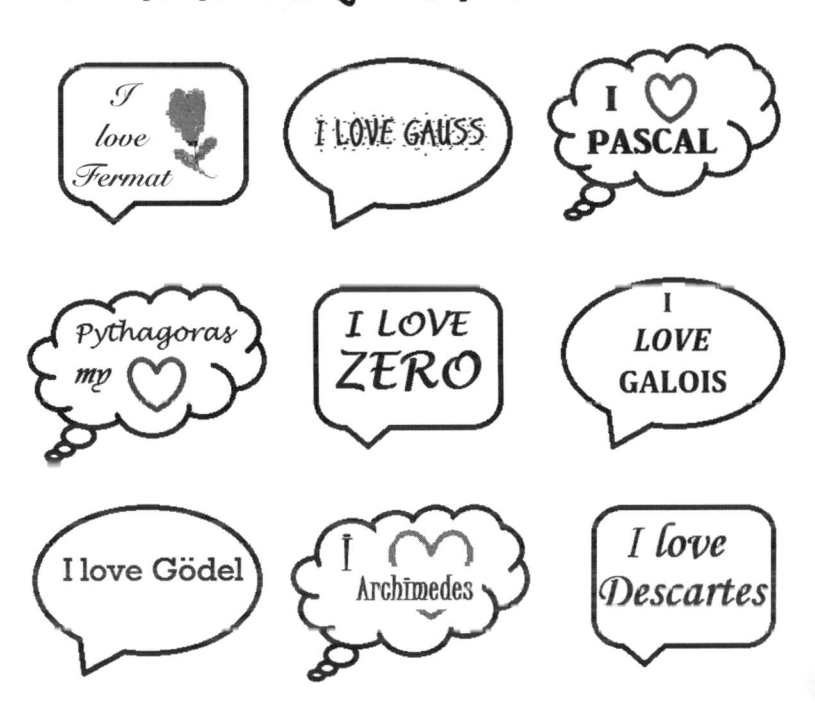

- I love Fermat
- I LOVE GAUSS
- I ♥ PASCAL
- Pythagoras my ♥
- I LOVE ZERO
- I LOVE GALOIS
- I love Gödel
- I ♥ Archimedes
- I love Descartes

More Mathematical Quickies & Trickies

Answers/Hints/Solutions

Mathematical Quickies & Trickies 1 (p. 4)

1. Pages 7, 8 and 14.

2. Not 1 : 4, but 1 : 3.
   Cut 2 circles with these ratios and act it out to convince yourself.

3. Five frogs.

4. Six o'clock only.

5. 7.
   Observe that the last digit of the powers of 7 ends with the pattern:
   7, 9, 3, 1, 7, 9, 3, 1, ….
   The last digit repeats in cycles of four, that is, 7^{4k} ends in 1.
   Since $7777 = (4 \times 1944) + 1$, the last digit of 7^{7777} is 7.

6. 62.5.

7. 1 : 1 (The rectangle should be a square.)
   Let the length and width be $\left(\frac{x}{2} + e\right)$ and $\left(\frac{x}{2} - e\right)$, where $0 < e < \frac{x}{2}$.
   Then $\left(\frac{x}{2} + e\right) + \left(\frac{x}{2} - e\right) = x$
   Now $\left(\frac{x}{2} + e\right)\left(\frac{x}{2} - e\right) = \frac{x^2}{4} - e^2$
   Since $-e^2 \leq 0$, the product $\frac{x^2}{4} - e^2$ is maximum when $e = 0$, that is, maximum value = $\frac{x^2}{4}$.
   Then x should be divided into two equal parts.
   Hence, the ratio is $\frac{x}{2} : \frac{x}{2} = 1 : 1$.

8. Eight members.
   Hint: Consider simple cases with fewer number of people, and observe any pattern from the corresponding number of handshakes.
   Show that for n number of people, there are $(n-1) + (n-2) + \ldots + 3 + 2 + 1$ handshakes.
   Then $\frac{n(n-1)}{2} = 28$
   Solve n.

9. $xy = yz = zw = wx = 1$
   Since $xy = yz$, $x = z$
   Since $yz = zw$, $y = w$
   Also, $x = \frac{1}{y}$, $y = \frac{1}{z}$, $z = \frac{1}{w}$, $w = \frac{1}{x}$.
   $xyzw = \frac{1}{y} \times \frac{1}{z} \times \frac{1}{w} \times \frac{1}{x} = \frac{1}{xyzw}$
   $(xyzw)^2 = 1$
   $xyzw = \pm 1$
   If $xyzw = -1$, and $xy = yz = zw = wx = 1$, no value of $x, y, z,$ and w exists.
   If $xyzw = 1$, and $xy = yz = zw = wx = 1$, then
   $x = 1, y = 1, z = 1, w = 1$, or
   $x = -1, y = -1, z = -1, w = -1$, or
   $x = n, y = \frac{1}{n}, z = n, w = \frac{1}{n}$.

10. 15 mathium in 15 pots.
    There are more than one way to solve the rate problem, two of which are given below.
    $1\frac{1}{2}$ men take $1\frac{1}{2}$ days to plant $1\frac{1}{2}$ mathium in $1\frac{1}{2}$ pots.
    $\frac{1}{2}$ man takes $1\frac{1}{2}$ days to plant $\frac{1}{2}$ mathium in $\frac{1}{2}$ pot.
    1 man takes $1\frac{1}{2}$ days to plant 1 mathium in 1 pot.
    1 man takes 9 days to plant $(1 \times \frac{2}{3} \times 9)$
    = 6 mathium in 6 pots.
    $1\frac{1}{2}$ men takes 9 days to plant 9 mathium in 9 pots.
    $2\frac{1}{2}$ men take 9 days to plant $(6 + 9) = 15$ mathium in 15 pots.
    Alternatively,
    $1\frac{1}{2}$ men take $1\frac{1}{2}$ days to plant $1\frac{1}{2}$ mathium in $1\frac{1}{2}$ pots.
    $1\frac{1}{2}$ men take 1 day to plant 1 mathium in 1 pot.
    $1\frac{1}{2}$ men take 9 days to plant 9 mathium in 9 pots.
    $2\frac{1}{2}$ men take 9 days to plant $9 \times \frac{2}{3} \times \frac{5}{2}$
    = 15 mathium in $9 \times \frac{2}{3} \times \frac{5}{2} = 15$ pots.

Mathematical Quickies & Trickies 2 (p. 12)

1. 5 cassettes, 74 minutes.

   Playing time = 92 + 88 + 95 + 101
   = 376 > 4 × 90

   Thus, 5 cassettes are needed, with the fifth one with playing time of (376 − 360) = 16 minutes.

   Hence, the fifth cassette will have (90 − 16), or 74 minutes of blank tape.

2. They refer to the number of seconds in a minute, minutes in an hour, hours in a day, days in a week, and weeks in a year.

3. Yes, Abram will arrive at 12 noon.
   The journey is 100 km and takes 2 hours.

4. 40 feet.

   Let r is the radius of the circular table.
   Then $r^2 = (r-8)^2 + (r-4)^2$
   $r^2 = r^2 - 16r + 64 + r^2 - 8r + 16$
   $r^2 - 24r + 80 = 0$
   $(r-4)(r-20) = 0$
   $r = 4$ or $r = 20$
   But $r \neq 4$ as r is too small.
   Therefore the radius, r, must be 20 feet.
   Hence the diameter of the circular table is 40 feet.

5. Acid before = Acid after
   Water after = Water before + Water added
   Total volume before = Acid before + Water before
   Total volume after = Acid after + Water after

   Method 1
   Water is 0% acid.
   If x cm³ of water are added, then
   $0.9 \times 1000 + 0 \times x = 0.8 \times (1000 + x)$
   $900 = 800 + 0.8x$
   $x = \frac{100}{0.8} = 125$
   So 125 cm³ of water must be added to yield an 80% acid mixture.

   Method 2
   Let x cm³ of water be added.
   $\frac{100 + x}{900} = \frac{20}{80} = \frac{1}{4}$
   $400 + 4x = 900$
   $4x = 500$
   $x = 125$

6.

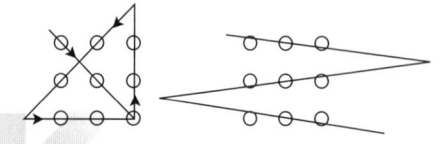

7. 1234567.

   Given: $\frac{1234567}{1234568^2 - (1234567 \times 1234569)}$

   Let $x = 1234567$
   Then $\frac{x}{(x+1)^2 - [x \times (x+2)]} = \frac{x}{x^2 + 2x + 1 - x^2 - 2x}$
   $= \frac{x}{1}$
   $= x$

   Therefore $\frac{1234567}{1234568^2 - (1234567 \times 1234569)}$
   $= 1234567$

8. 71.125%.

   No invention can save 100% on fuel, since energy cannot rise from nothing.

   The correct calculation is not 30% + 45% + 25% = 100%, but: 100% − (100% − 30%)(100% − 45%)(100% − 25%) = 100% − (70% × 55% × 75%)
   = 71.25% (assuming the three inventions are independent in effect.)

9. The third child is Robin.

10. 10 inches.
    The diameter of the log is not $\frac{45}{\pi}$. The distance from one knothole to another cut of the same knothole is about two thirds the width of the whole plywood sheet, or 30 inches. The diameter of the log is $\frac{30}{\pi} \approx 10$ inches.

Mathematical Quickies & Trickies 3 (p. 19)

1. 12,345,678,987,654,321.

   999,999,999 × 12,345,679
   = (12,345,679 × 10⁹) − 12,345,679

   12,345,679,000,000,000
   −   12,345,679
   ―――――――――――――――
   12,345,678,987,654,321

2. 7:15 p.m.

   2:00 on Sunday to 8:00 on Monday = 24 h + 6h
   = 30 h
   6 h → 9 min
   30 h → $\frac{9}{6} \times 30$ = 45 min
   8:00 − 45 min = 7:15

182   More Mathematical Quickies & Trickies − Answers/Hints/Solutions

3. 1053.

   A B C ... Y Z
   1 2 3 ... 25 26
   Total value of all the letters
   $= 3 \times (1 + 2 + 3 + ... + 26)$
   $= 3 \times \frac{26 \times 27}{2}$
   $= 1053$

4. C.

   If we assign A = 1, B = 2, C = 3, ..., then
   S T O P
   19 + 20 + 15 + 16 = 70
   S T A R T
   19 + 20 + 1 + 18 + 20 = 78

5. 9 cm.

   One possible path is as shown:
   ABCDEFGHAD
   Any other path covers 9 cm
   or less.

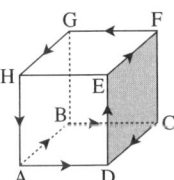

6. Remainder = 1.

   Since $10^{99} = 999...999 + 1$, $10^{99} \div 9$ leaves a remainder of 1.

7. 13 steps.

   On counting the last 5 steps, he reaches the top floor. Thus, when he has counted to 40, he has reached a point 5 stairs below the landing floor. Up to this point each five he has counted corresponds to a net movement up the staircase of one stair. Therefore, before this point, he has come up 8 stairs. Hence, there must be 13 steps altogether before his landing.

8. 30.

   Any 3-digit number will appear as one of the following:
   1?1, 2?2, 5?5, 6?9, 8?8, 9?6.

   For any of the 6 possible positions, there are five ways to fill in "?": 0, 1, 2, 5, or 8.

   Therefore, there are $6 \times 5 = 30$ such numbers between 100 and 1000.

9. $3.00.

   Note that we need to buy an even number of exercise books.

   If we buy two exercise books, one pen and one pencil, we will spend $1.24 ($2 \times 0.37 + 0.22 + 0.28$). That implies that $1.00 cannot be the least number of dollars.

The difference of $2 and $1.24 is $0.76. There is no way we can spend 76 cents. So, $2 is out.

Let us try $1.76 for a total of $3.
If we buy four exercise books, that would leave us $(1.76 − $1.48) = 0.28, the price of a pen. Alternatively, we may also spend $1.76 by buying eight additional pencils.

Hence, the least whole number of dollars is three.

10. 7 members.

    Hint: Show that at least 6 members like algebra and geometry, and at least one member likes algebra and calculus.

Mathematical Quickies & Trickies 4 (p. 24)

1. 0.

   Since the prime number between 1 and 1999 include both 2 and 5, their product is divisible by 10. Thus, the remainder is 0.

2. 1999.

   1 million = 1000^2 and 9 million = 3000^2.
   The perfect squares between 1000^2 and 3000^2 are:
   $1001^2, 1002^2, ..., 2999^2$.

   Thus, there are 1999 such squares.

3. 5.

   $\frac{1+5}{4+5} = \frac{6}{9} = \frac{2}{3}$

4. 17.

5. Only one day a week.

   If Samuel chooses the first sock at random, there remain 7 socks for him in the drawer.

   Only one of these will match his first sock. Therefore, he has a one in seven chance of picking the right one. Hence, he will wear a matching pair of socks on average only one day a week.

6. 2720th.

   The 776th "*t*" occurs at the $\frac{776}{2} \times 7 = 2716$th letter of the pattern.

   The 777th "*t*" occurs at the $(2716 + 4) = 2720$th letter of the pattern.

7. 48 points.

   One needs 4 points to win a game.

   Therefore, to win 2 sets, one needs $2 \times 6 \times 4 = 48$ points.

8. Zero.

   Since $17 + 42 = 559$, what results is a very skinny triangle. Its area is zero.

9. $25^2 = 625$.

   Since both HE and SHE end in the same digit, E can take one the following: 0, 1, 5, or 6.

   Since SHE is less than 1000, HE cannot exceed 32 ($\sqrt{1000} \approx 31.6$). That leaves 9 possibilities: 10, 15, 16, 20, 21, 25, 26, 30, and 31.

   The only possibility turns out to be HE = 25, so that $25^2 = 625 =$ SHE.

10. 249 zeros.

    The number of 5's in $1 \times 2 \times 3 \times \ldots \times 1000$ is $200 + 40 + 8 + 1 = 249$.

    There are enough of 2's to match each 5 to get a 10.

    Hence, $1 \times 2 \times 3 \times \ldots \times 1000$ ends in 249 zeros.

Mathematical Quickies & Trickies 5 (p. 31)

1. Only was going to St. Ives.

2. 10,011,010.

3. $\frac{1}{7}$.

   Hint: Let the sequence of numbers be denoted by $T_1, T_2, T_3, T_4, \ldots$.
   Show that $T_2 = x$.

   Observe that the sequence of numbers alternates between x and $\frac{1-x}{1+x}$.

   $T_{100} = T_{\text{even no.}}$

4. 4.

   Observe that the last digit of the powers of 2 exhibits a pattern: 2, 4, 8, 6, 2, 4, 8, 6, …

   $2002 = 4 \times 500 + 2$

   Thus the last digit of 2^{2002} is 4.

   Hence, the remainder when 2^{2002} is divided by 10 is 4.

5. $3\frac{1}{2}$.

   $(x + y) = x^2 + 2xy + y^2 = 12 + 1$
   $(x^2 + y^2) = x^4 + 2x^2y^2 + y^4 = 2^2 = 4$
   $x^2y^2 = (xy)^2$
   $2xy = (x + y)^2 - (x + y^2)$
   $\quad = 1 - 2 = -1$
   $xy = -\frac{1}{2}$
   $x^4 + y^4 = (x^2 + y^2) - 2x^2y^2$
   $\quad = 4 - 2\left(-\frac{1}{2}\right)^2$
   $\quad = 3\frac{1}{2}$

6. $10\frac{1}{2}$ hours.

   David does $\frac{2}{3}$ of the work in 7 hours.

   Therefore, the whole job will take David $7 \times \frac{3}{2}$
   $= 10\frac{1}{2}$ hours.

7. 34 years old.

   The only square of an integer between 1700 and 1800 is $1764 = 42^2$.

   Therefore, the woman was born in $(1764 - 42) = 1722$.

   Hence, in 1756, she was $1756 - 1722 = 34$ years old.

8. 4.

   $\frac{1}{p} + \frac{1}{q} = \frac{p + q}{pq} = \frac{1}{pq}$

   Since $2pq \leq p^2 + q^2$ and $p^2 + q^2 = (p = q)^2 - 2pq$, we have $2pq \leq (p + q)^2 - 2pq$
   $4pq \leq (p + q)^2 = 1$
   $pq \leq \frac{1}{4}$
   $\frac{1}{pq} \geq 4$

9. $56\frac{1}{4}\%$.

   Hint: $(1.25x)^2$.

10. 2:04 p.m.

    The train must travel a distance of 2 km (length of tunnel + length of train).

    At 30 km/h, the train travels 1 km in 2 minutes.

    Thus the train travels 2 km in 4 minutes.

    Hence the rear of the train comes out of the tunnel at 2:04 p.m.

Mathematical Quickies & Trickies 6 (p. 37)

1. 54.

Let the 2-digit number be ab, where $a > b$.

Then $\dfrac{ab - ba}{ba} \times 100 = 20$

$\dfrac{(10a + b) - (10b + a)}{10b + a} \times 100 = 20$

$\dfrac{9a - 9b}{10b + a} = \dfrac{1}{5}$

$45a - 45b = 10b + a$

$44a = 55b$

$4a = 5b$

For $4a = 5b$, the integer solutions are $a = 5$ and $b = 4$.

Thus the number is 54.

Check: $\dfrac{54 - 45}{45} \times 100 = 20$.

2. 140°.

3. 262,273.

We need to find the LCM of 1243 and 23,843.

HCF (1243, 23,843) = 113

Since HCF$(x, y) \times$ LCM$(x, y) = xy$ for any two numbers x and y,

LCM (1243, 23,843) = $\dfrac{1243 \times 23{,}843}{113} = 262{,}273$

Therefore, the society is made up of at least 262,273 members.

4. John has $35 and Jane has $25.

When John gives Jane $5, both have the same amount of money. This means that John originally has ($2 × 5) = $10 more than Jane.

When Jane gives John $5, John will have another $10 more than Jane. In other words, John will now have $20 more than Jane.

This extra $20 enables John to have twice as much money as Jane. This means that Jane must have herself $20 left after giving $5 to John. In other words, Jane must have originally $(20 + 5) = $25.

And John, who has $10 more than Jane, must have $25 + 10 = $35.

Note: A visual method to solve the problem would be to use the model method. Try it!

5. 40 km/h.

Hint: Distance downstream = distance upstream

(Rate of boat + rate of current) × time downstream = (Rate of boat − rate of current) × time upstream

6. $\dfrac{12}{25}$ hour.

In 12 hours, the first pipe fills 12 tanks,
the second pipe fills 6 tanks,
the third pipe fills 4 tanks, and
the fourth pipe fills 3 tanks.

Therefore, 25 tanks can be filled by all four pipes in 12 hours.

Hence, one tank can be filled by all four pipes in $\dfrac{12}{25}$ hour.

7. 1296 rectangles.

The top row contains $8 + 7 + 6 + 5 + 4 + 3 + 2 + 1 = 36$ rectangles ($1 \times n$). There are 8 such rows. Thus, there are (8×36) rectangles ($1 \times n$).

Similarly, on the top pair of rows there are 36 rectangles ($2 \times n$), and there are 7 such pairs. Thus, there are (7×36) rectangles ($2 \times n$).

Continuing this process, we have (6×36) rectangles ($3 \times n$), and so on.

Total number of rectangles = $(8 \times 36) + (7 \times 36) + (6 \times 36) + \cdots + (1 \times 36)$
$= 36 \times (8 + 7 + 6 + 5 + 4 + 3 + 2 + 1)$
$= 36 \times 36$
$= 1296$

8. The prolem cannot be solved, as we are not given the sample size.

9. To show that $\sqrt{1 + \sqrt{1 + \sqrt{1 + \sqrt{1 + \ldots}}}}$

$= \dfrac{1}{\sqrt{1 + \sqrt{1 + \sqrt{1 + \sqrt{1 + \ldots}}}}}$

Let $y = \sqrt{1 + \sqrt{1 + \sqrt{1 + \sqrt{1 + \ldots}}}}$

Then $y = \sqrt{1 + y}$

$y^2 = 1 + y$

$y^2 - y - 1 = 0$

$y = \dfrac{1 \pm \sqrt{1 - 4 \cdot 1 \cdot (-1)}}{2 \times 1}$

$y = \dfrac{1 \pm \sqrt{5}}{2}$

Since $y > 0$, $y = \dfrac{1 + \sqrt{5}}{2}$

Let $x = \sqrt{1 + \sqrt{1 + \sqrt{1 + \sqrt{1 - \ldots}}}}$

Then $x = \sqrt{1 - x}$

$x^2 = 1 - x$

$x^2 + x - 1 = 0$

$x = \dfrac{-1 \pm \sqrt{1 - 4 \cdot 1 \cdot (-1)}}{2 \cdot 1}$

$x = \dfrac{-1 \pm \sqrt{5}}{2}$

Since $x > 0$, $x = \dfrac{-1 + \sqrt{5}}{2}$

Now $xy = \dfrac{1 + \sqrt{5}}{2} \cdot \dfrac{-1 + \sqrt{5}}{2}$

$xy = \dfrac{\sqrt{5} + 1}{2} \cdot \dfrac{\sqrt{5} - 1}{2} = 1$

Therefore,

$\sqrt{1 - \sqrt{1 - \sqrt{1 - \sqrt{1 - \cdots}}}} \cdot \sqrt{1 + \sqrt{1 + \sqrt{1 + \sqrt{1 + \cdots}}}}$
$= 1$

Hence,

$\sqrt{1 + \sqrt{1 + \sqrt{1 + \sqrt{1 + \cdots}}}} = \dfrac{1}{\sqrt{1 - \sqrt{1 - \sqrt{1 - \sqrt{1 - \cdots}}}}}$

10. $x = 2$, $y = 4$, $z = 6$.

   Hint: Add the three equations simultaneously.

Mathematical Quickies & Trickies 7 (p. 44)

1. 5 girls.

   The boys and girls must alternate, so that there are as many girls as boys.

2. 204 squares.

   There are 8^2 squares (1×1);
   7^2 squares (2×2);
   6^2 squares (3×3), and so on.

   Therefore there are altogether $(8^2 + 7^2 + 6^2 + 5^2 + 4^2 + 3^2 + 2^2 + 1^2) = 204$.

   [*Note:* A useful formula for the sum of the squares is: $1^2 + 2^2 + 3^2 + \ldots + n^2 = \dfrac{1}{6}n(n+1)(2n+1)$.]

3. (i) $\dfrac{2}{3}$, (ii) $-\dfrac{2}{9}$, (iii) $-\dfrac{10}{27}$.

   (i) $\dfrac{1}{x} + \dfrac{1}{y} = \dfrac{y + x}{xy} = \dfrac{2}{3}$

   (ii) $\dfrac{1}{x^2} + \dfrac{1}{y^2} = \dfrac{y^2 + x^2}{x^2 y^2} = \dfrac{(x+y)^2 - 2xy}{(xy)^2} = \dfrac{2^2 - 2(3)}{3^2}$
   $= \dfrac{-2}{9}$

   (iii) $\dfrac{1}{x^3} + \dfrac{1}{y^3} = \dfrac{y^3 + x^3}{x^3 y^3} = \dfrac{(x+y)^3 - 3x^2y - 3xy^2}{(xy)^3}$
   $= \dfrac{2^3 - 3(3)(2)}{3^3} = \dfrac{-10}{27}$

4. 25 or 36.

   Let the 2-digit number be ab.
   Then $ab = b^2$
   $10a + b = b^2$
   $b^2 - b = 10a$
   $b(b - 1) = 10a$

   We test for integer solutions for $0 < a < 9$.
   When $a = 1$, $b(b-1) = 10$.
   No solution exists.
   When $a = 2$, $b(b-1) = 20$.
   $(b-1) = 4$, $b = 5$
   Then $ab = 25$, and $5^2 = 25$
   When $a = 3$, $b(b-1) = 30$.
   $(b-1) = 5$, $b = 6$
   Then $ab = 36$, and $6^2 = 36$
   No solution exists for $a = 4, 5, \ldots, 9$.
   Hence, the 2-digit numbers are 25 and 36.

5. 6893 s.

Numbers	Number of digits	Time (s)
1–9	9	9
10–99	90×2	180
100–999	900×3	2700
1000–2000	1001×4	4004
Total		6893

6. Mr. Yan made his statement on January 1, and he was born on December 31. He will turn 43 at the end of the next calendar year.

7. Saturday.

   What is today if 3 days from now will be Wednesday? (Sunday)
   If today is Sunday, what is the day before yesterday? (Friday)
   What day follows Friday? (Saturday)

8. n^2.

   Because there is only one winner, there must be $(n^2 + 1) - 1 = n^2$ losers.
   Since there is one loser per game, there must be n^2 games.

9. The inner circle is smaller.

   Area of the outer circle $= \pi \times \left(\dfrac{6}{2}\right)^2 = 9\pi$

   Area of the inner circle $= \pi \times \left(\dfrac{4}{2}\right)^2 = 4\pi$

   Area of the space between the circles $= 9\pi - 4\pi$
   $= 5\pi$

   Therefore, the inner circle is smaller.

10. 3^{20} is larger
    $3^{20} = 3^{2 \times 10} = (3^2)^{10} = 9^{10}$
    $2^{30} = 2^{3 \times 10} = (2^3)^{10} = 8^{10}$
    Since $9^{10} > 8^{10}$, $3^{20} > 2^{10}$.

Mathematical Quickies & Trickies 8 (p. 50)

1. 9 and 17.

   Challenge: Use a graphical method to solve the question.

2. Since the first digit of the code cannot be 0, 2 or 5, we are left with 7 possible numbers for the first digit. However, all ten digits can be used for the second, third and fourth numbers.

   Hence there are $7 \times 10 \times 10 \times 10 = 7000$ possible different codes.

3. 2112.
   $2000 \div 132 \approx 15.15\ldots$
   $132 \times 16 = 2112$

4. 70 routes.

5. Monday.

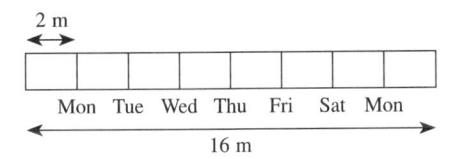

   ```
   2 m
   ←→
   ┌──┬──┬──┬──┬──┬──┬──┬──┐
   │  │  │  │  │  │  │  │  │
   └──┴──┴──┴──┴──┴──┴──┴──┘
   Mon Tue Wed Thu Fri Sat Mon
   ←──────── 16 m ────────→
   ```

   She needs only 7 cuts to get 8 parts of 2 m each.

   She will make her cuts on Monday (1st cut), Tuesday (2nd cut), Wednesday (3rd cut), Thursday (4th cut), Friday (5th cut), Saturday (6th cut) and Monday (7th cut), since she rested on Sunday.

6. $58 \times 3 - 174 = 29 \times 6$.
   Are there other solutions?

7. 12 h or 720 min.

8. 15 routes $(= 1 + 2 + 3 + 4 + 5)$

   If Joe takes path 1, he can get back home with any paths from 2 to 6.

   Similarly, if he takes path 2, he can get back home with any road from 3 to 6; if he takes path 3, he can get back home with any path from 4 to 6; and so on. And if he takes path 6, he can get back only by path 7.

   The total number of routes he can choose is:
   $5 + 4 + 3 + 2 + 1 = \frac{1}{2} \times 5 \times 6 = 15$

   So Joe can use 15 different routes.

9. (a) $1 \times 2 \times 3 \times 4 \times 5 \times 6 \times 5 \times 4 \times 3 \times 2 \times 1$
       $= 86,400$ ways

   (b) The number of routes to each letter is given by:

   ```
                      1
                   1     1
                1     2     1
             1     3     3     1
          1     4     6     4     1
       1     5    10    10     5     1
          6    15    20    15     6
             21    35    35    21
                56    70    56
                  126   126
                     252
   ```

   Hence the number of ways is 252.

10. 381,654,729
    $3 = 1 \times 3$ $381,654 = 6 \times 63,609$
    $38 = 2 \times 19$ $3,816,547 = 7 \times 545,221$
    $381 = 3 \times 127$ $38,165,472 = 8 \times 4,770,684$
    $3816 = 4 \times 954$ $381,654,729 = 9 \times 42,406,081$
    $38,165 = 5 \times 7633$

Mathematical Quickies & Trickies 9 (p. 56)

1. (a) 10,000.   (b) 10,000.
   (a) $101^2 - 201 = 101^2 - 202 + 1$
                   $= 101^2 - 2(101)(1) + 1^2$
                   $= (101 - 1)^2$
                   $= 100^2$
                   $= 10,000$
   (b) $99^2 + 200 - 1 = 99^2 - 1^2 + 200$
                      $= (99 - 1)(99 + 1) + 200$
                      $= 98 \times 100 + 200$
                      $= 100 \times (98 + 2)$
                      $= 100 \times 100$
                      $= 10,000$

2. $21.
   Difference in savings = $(72 - 48) = 24.
   Each week, Sam saves $3 more than Bob.
   Therefore, a difference of $24 will take the two boys $\frac{24}{3} = 8$ weeks to save.
   Bob saves $48 in 8 weeks.
   Bob spends $(15 \times 8) = 120 on food in 8 weeks.
   Therefore, Bob's pocket money for 8 weeks was $(120 + 48) = 168.
   Hence, Bob's weekly pocket money $= \$\frac{168}{8} = \$21$

3. 154 and 155.

   If x and $x + 1$ are the page numbers, then $x(x + 1) = 23{,}870$
   Now $x(x + 1) \approx x^2$
   So $x^2 \approx 23{,}870$
   $x \approx \sqrt{23{,}870} = 154.45$
   Take $154 \times 155 = 23{,}870$

   Hence the page numbers are 154 and 155.

4. $1560.

   20 pieces of $10 notes = $200
   Number of $2 notes = 100

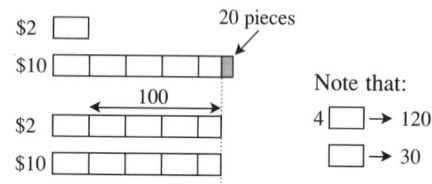

   Note that:
   $4 \square \to 120$
   $\square \to 30$

   Paul had $(\$2 \times 30) + (\$10 \times 150) = \$1560$

5. 60%.

   24 days are needed to consume the food.
   1 day will consume $\frac{1}{24}$ of the food.
   Since the food is to last for 40 days, therefore, $\frac{1}{40}$ of the food has to been consumed in one day.
   Percentage decrease in food consumption
   $= \dfrac{\frac{1}{24} - \frac{1}{40}}{\frac{1}{24}} \times 100 = 40\%$

   Hence, each recruit's ration must be decreased by 60%.

6. (a) 24 km², (b) 600 cm².

   Scale = 1 : 200,000
   1 cm ≡ 200,000 cm
   1 cm² = 200,000² cm²
   Actual area = $(6 \times 200{,}000^2)$ cm²
   $= \dfrac{6 \times 200{,}000^2}{100 \times 100}$ m²
   $= \dfrac{6 \times 200{,}000^2}{100 \times 100 \times 1000 \times 1000}$ km²
   = 24 km²

   (c) Scale 1 : 20,000
   Area of region $= \dfrac{6 \times 200{,}000^2}{20{,}000^2}$ cm²
   = 600 cm²

7. 6 pages and 10 stickers.

   An intuitive approach is as follows:

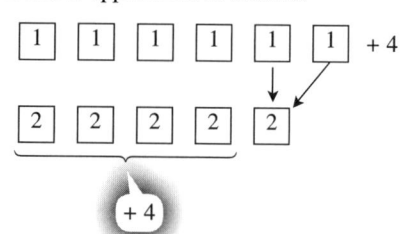

8. 127 eggs.

   Algebraically,
   If N represents the number of eggs.
   Observe that the number of eggs sold to each customer follows a pattern:
   $\frac{1}{2}(N + 1), \frac{1}{2^2}(N + 1), \frac{1}{2^3}(N + 1), \ldots$
   $N = \frac{1}{2}(N + 1) + \frac{1}{2^2}(N + 1) + \frac{1}{2^3}(N + 1)$
   $+ \cdots + \frac{1}{2^7}(N + 1)$
   Simplifying, $N = \frac{127}{128}(N + 1)$
   $N = 127$

9. 4000 toys; 15 days.

   Hint: An extra 40 toys are made each day.
   10 days are needed to make up the 400 toys.
   The extra 200 toys must have come from
   $\frac{200}{40} = 5$ days.

10. $\dfrac{xy}{x + y}$.

    In 1 hour the boys travel $\left(\dfrac{1}{x} + \dfrac{1}{y}\right) = \dfrac{y + x}{xy}$ of the distance AB.
    $\dfrac{y + x}{xy}$ of the distance AB is covered in 1 hour.
    The whole distance AB is covered in $\dfrac{xy}{x + y}$ hour.

Mathematical Quickies & Trickies 10 (p. 64)

1. (a) 1.
   $300^2 - 301 \times 299 = 300^2 - (300 + 1)(300 - 1)$
   $= 300^2 - (300^2 - 1)$
   $= 300^2 - 300^2 + 1$
   $= 1$

(b) 1,000,000.

$$999^2 + 1999 = 999^2 + (1000 + 999)$$
$$= 999 \times (999 + 1) + 1000$$
$$= (999 \times 1000) + 1000$$
$$= 1000 \times (999 + 1)$$
$$= 1000 \times 1000$$
$$= 1,000,000$$

2. 2^{2520}.

   LCM (2, 3, 4, 5, 6, 7, 8, 9) = 2520

   The smallest such number is 2^{2520} because we require a value greater than 1.

3. 22.

   $n = 2747q + 79$
   $= 67 \times (41q) + 79$
   $= 67 \times (41q + 67 + 22)$
   $= 67 \times (41q + 1) + 22$

   Therefore, the remainder when n is divided by 67 is 22.

4. 15 km.

   The trains' relative speed to each other is 80 km/h. So it takes $\frac{1}{4}$ hour to cover the 20 km to collision.

   In $\frac{1}{4}$ hour the fly has traveled 15 km.

5. (a) 3 a.m. Sunday
   (b) 10 p.m. Saturday
   (c) 1 a.m. Sunday
   (d) 8 p.m. Saturday
   (e) 2 a.m. Sunday

6. $1\frac{7}{8}$ kg.

   Hint: $L + B = \frac{3}{2}$
   $L = \frac{3}{4} + B$

   Weight of bottle when full = $B + \frac{4}{3}L$.

   Or, use a model method to solve the problem.

7. Current speed = 2 km/h; Boat speed = 6 km/h.

   Let V_c km/h be the speed of the current, and V_b the speed of the boat in still water.

   Upstream: $\frac{6}{V_b - V_c} = 1\frac{1}{2} = \frac{3}{2}$
   $3V_b - 3V_c = 12$ (1)

   Downstream: $\frac{6}{V_b + V_c} = \frac{3}{4}$
   $3V_b + 3V_c = 24$ (2)

   Solve equations (1) and (2) simultaneously.
   Or, use an intuitive method to solve the problem.

8. Ruth: $\frac{4}{100} \times 120 = 4.8$ units
   Job: $2 \times 4.8 = 9.6$ units

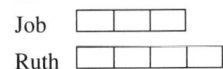

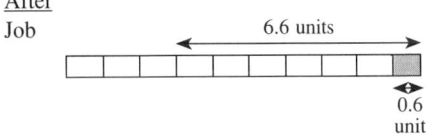

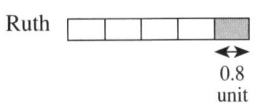

   Before
   Job
   Ruth

   After
   Job — 6.6 units, 0.6 unit
   Ruth — 0.8 unit

   Job must save $\frac{6.6}{3} \times 100\% = 220\%$.

9. 9 m/min and 36 m/min.

   Let x m/min and $4x$ m/min be the speed of the first cyclist and of the second cyclist, respectively.

   Then $37x + [999 + 37x] = 37 \times (4x)$

   Simplifying, $x = 9$

10. 20, 12, 4, 64.

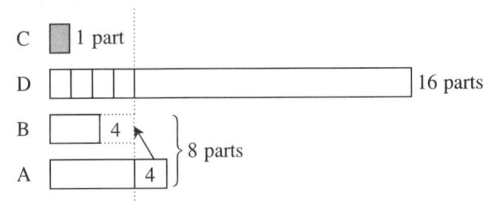

    C — 1 part
    D — 16 parts
    B — 4
    A — 4
    } 8 parts

    Observe A and B total 8 parts.
    25 parts → 100
    1 part → 4
    The numbers are 20, 12, 4, and 64.
    Check: $20 - 4 = 12 + 4 = 4 \times 4 = 64 \div 4$.

Mathematical Quickies & Trickies 11 (p. 70)

1. $(a + b + c)^2 = 1^2$
   $a^2 + b^2 + c^2 + 2ab + 2bc + 2ac = 1$
   $ab + bc + ac = \frac{1}{2} - \frac{a^2 + b^2 + c^2}{2} < \frac{1}{2}$

2. -8.

   Completing the square, we have
   $x^2 + y^2 - 8x + 6y + 17$
   $= x^2 - 8x + 16 + y^2 + 6y + 9 - 8$
   $= (x - 4)^2 + (y + 3)^2 - 8$

   The minimum occurs at -8 when $x = 4$ and $y = -3$.

3. Jacob.

   Esau covers the last $\frac{3}{12}$ of the race in 3 seconds.

   Jacob covers the last $\frac{4}{12}$ of the race in 4 seconds.

   Both average $\frac{1}{12}$ of the race per second for the respective distance.

   Since Jacob maintains this average rate over a greater distance, he must achieve it first. Hence, Jacob is the winner.

4. $90\frac{10}{11}$ km/h.

5. 35.

6. 80 steps.

   If s is the number of steps, and r is the rate of the escalator (steps per second),
   then $s - 20 = 20r$ (1)
         $s - 32 = 16r$ (2)

   Solving equations (1) and (2) simultaneously, we have $s = 80$.

7. First simultaneously light one piece of fuse at both ends and the second piee at one end.

   As soon as the first place finishes burning, light the second piece at its other end.

   The second fuse will burn for 15 more seconds, which completes the 45 seconds.

8. $(x^2 + 2x + 2)(x^2 - 2x + 2)$.

   Complete the square:
   $x^4 + 4 = x^4 + 4x^2 + 4 - 4x^2$
   $= (x^2 + 2)^2 - (2x)^2$
   $= (x^2 + 2x + 2)(x^2 - 2x + 2)$

9. At every birth, regardless the history, a boy or a girl is equally likely, so one expects equal numbers of each.

10. $t = n + \frac{t}{144}$, for $n = 2, 3, 4, \ldots, 143$.

    Observe that if we push the minute hand around on a clock the hour hand follows at $\frac{1}{12}$ the speed.

    Suppose we placed two clocks, both reading 12:00, side by side, and pushed the minute hand of one and the hour hand of the other forward at the same rate.

    The minute hand of the second clock would sweep around $12 \times 12 = 144$ times as fast as the hour hand of the first, and therefore would catch up shortly after all the way around once.

At that instant it would indicate a time t (in hours) given by $t = 1 + \frac{t}{144}$, or $t = \frac{144}{143}$ hours = 1 hour and 25.2 seconds.

The other clock would read $\frac{1}{12}$ as much, or $\frac{12}{143}$ hour = 5 minutes and 2.1 seconds.

If their hour and minute hands were identical the two clocks would be indistinguishable.

The same thing would happen each time the fast minute hand overtook the slow hour hand, at times t satisfying $t = n + \frac{t}{144}$, for $n = 2, 3, 4, \cdots, 143$.

Note that the last time, when $n = 143$, both hands are once again pointing upward on both clocks.

Mathematical Quickies & Trickies 12 (p. 77)

1. 3^{30}.

   $2^{40} = (2^4)^{10} = (16)^{10} < (27)^{10} = (3^3)^{10} = 3^{30}$

   Thus 3^{30} is greater than 2^{40}.

2. 59.

   LCM (2, 3, 4, 5) − 1.

3. 241, or 2,199,023,255,552.

   The father has 2 children (1st generation), then $2 \times 2 = 2^2 = 4$ grandchildren (2nd generation), and $2 \times 2 \times 2 = 2^3 = 8$ great grandchildren (3rd generation).

   In the year 3000 (at the 41st generation), he will have 2^{41}, or 2,199,023,255,552, ever-so-great grandchildren. That is about 2.2 trillion new descendants.

4. Not 11, but 10 times.

5. $x^3 - y^3 - z^3 - 3x^2z + 3xz^2$
   $= (x^3 - 3x^2z + 3xz^2 - z^3) - y^3$
   $= (x - z)^3 - y^3$
   $= (x - z - y)[(x - z)^2 + (x - z)y + y^2]$
   $= (x - z - y)(x^2 + y^2 + z^2 + xy - 2xz - yz)$

6. Five 4¢ stamps, fifty 2¢ stamps and eight 10¢ stamps.

   Suppose Ruth bought x 4¢ stamps and y 10¢ stamps.

   Therefore, she spent $4x + 20x + 10y$ cents.

   Hence $24x + 10y = 200$,
       or $12x + 5y = 100$ (1)

190   More Mathematical Quickies & Trickies – Answers/Hints/Solutions

Rearranging (1): $12x = 100 - 5y$ (2)

Since 5 divides the RHS of (2), it must divide the LHS.

This is only possible if 5 divides x.

Hence, $x = 5k$ for some integer k.

Thus, $12(5k) = 100 - 5y$,

or $12k + y = 20$ (3)

Now, since $x > 0$, $k > 0$. Also, $y > 0$.

The only solution of (3) with $k > 0$ and $y > 0$ is $k = 1$, $y = 8$. (k cannot be greater than 1, since $12 \times 2 = 24 > 20$)

Hence, $x = 5k = 5$, and $y = 8$.

Ruth bought $x + 10x + y = 5 + 50 + 8 = 63$ stamps.

7. Observe that 0.249,999... is not approximately, but exactly equal to 0.25.

8. (a) 1; (b) 3; (c) 17,982.

   Hint:

   (a) Observe that the powers of 7 end with the pattern: 7, 9, 3, 1, 7, 9, 3, 1, ...
   The last digit repeats in cycles of four.

   (b) $27^1 = 27$; $27^2 = 729$; $27^3 = 19,683$;
   $27^4 = 931,441$; ...
   $27 = (4 \times 6) + 3$

   (c) $10^{1998} - 1 = 100 \ldots 000 - 1$
   $= 999 \ldots 999$

9. 5 cows, 1 pig and 94 sheep.

   Let x be the number of cows, y the number of pigs, and z the number of sheep.

   Then $1000x + 300y + 50z = 10,000$ (1)
   and $x + y + z = 100$ (2)

   From (1), $20x + 6y + z = 200$ (3)

   From (2), $z = 100 - x - y$ (4)

   Substitute (4) into (3), we have
   $20x + 6y + 100 - x - y = 200$
   $19x + 5y = 100$
   $5y = 100 - 19x$
   $y = \dfrac{100 - 19x}{5}$
   $= 20 - 3x - \dfrac{4x}{5}$

   Since $\dfrac{4x}{5}$ must be an integer, this means that x must be a multiple of 5.

   The lowest multiple of 5 is 5, so $y = \dfrac{100 - 19(5)}{5}$
   $= 1$, and $z = 100 - 5 - 1 = 94$.

   If x is a multiple greater than 5, y becomes negative.

Thus the problem has only one solution: $x = 5$, $y = 1$ and $z = 94$

Hence, Farmer Yan bought 5 cows, 1 pig and 94 sheep.

10. All three hands will be exactly together only at 12 o'clock.

    How many times do the hour and minute hands concide?

    This occurs only 10 times (not 12 times) between 12 noon and 12 midnight.

    The coincidence of the hands at 12 makes a total of 11 different times at which the two hands coincide.

    Using a similar reasoning, the second hand and the minute hand coincide at 59 different times.

    Thus, the hour hand and the minute hand coincide at 11 equal times, and the minute hand and the second hand coincide at 59 equal times.

    If x be is number of intervals between the first coincidences, and y is the number of intervals between the second coincidences.

    Then, x and y have a common factor z, there will be z spots where the two coincidences will occur simultaneously.

    Since 11 and 59 have no common factor, there cannot be a spot between 12 noon and 12 midnight when both coincidences occur at the same time.

    In other words, the three hands are exactly together only at 12 o'clock.

Mathematical Quickies & Trickies 13 (p. 86)

1. 57.

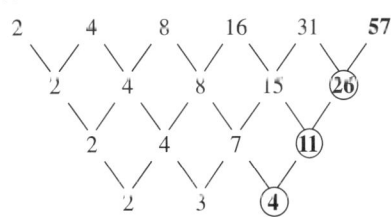

2. 21 seconds.

   Let Betty take x seconds, and l be the length of the track.

   Then $\dfrac{l}{40}(15) + \dfrac{l}{x}(15) = 1$

   $x = 24$

3. 0.

   Hint:

x	0	1	2	3	4	5	6	7	8	9
Units digit of x^3	0	1	8	7	4	5	6	3	2	9

4. 15 km/h.

   If the man runs away from the train, he will have one third of the length of the bridge to run after the train reaches the bridge.

   Thus, his speed is one third that of the train, or 15 km/h.

   Algebraically,

   Let the bridge be x km long, the train be y km from the bridge, and the speed of the man be v km/h.

   Then $\dfrac{\frac{x}{3}}{v} = \dfrac{y}{45}$ and $\dfrac{\frac{2x}{3}}{v} = \dfrac{x+y}{45}$

   Eliminating v, show that $x = y$.
   Then show that $v = 15$.

5. 169 cm².

   $xy + y = y^2 + 13$

   $x = \dfrac{13}{y} + y - 1$

   Observe that, since x is an integer, $y = 1$ or 13.
   If $y = 1$, $x = 13$
   If $y = 13$, $x = 1$
   Maximum area = $13 \times 13 = 169$

6. 8.

   $667 = 23 \times 29 = 1 \times 667$

   Thus, there are four positive divisors.

   Since y needn't be positive, therefore, we have four negative divisors for 667.

   Hence, there are altogether 8 divisors for 667.

7. Let x and $8\sqrt{2} - x$ be the length and width of the rectangle, respectively.

   If d is the diagonal of the rectangle, then, by Pythagoras' Theorem, $d^2 = x^2 + (8\sqrt{2} - x)^2$
   $d^2 = 2x^2 - 16\sqrt{2}x + 128$

   By completing the square, we have
   $d^2 = 2[(x - 4\sqrt{2})^2 + 32]$
   d^2 is minimum when $x = 4\sqrt{2}$.

   Therefore, the minimum value of d is
   $\sqrt{2(0^2 + 32)} = 8$.

8. 2.

   $(x + a)(x + b)(x + c) + 5 = 0$
   $x = 1$ is a solution implies that
   $(1 + a)(1 + b)(1 + c) = -5$
   Since a, b, and c are integers, $(1 + a)$, $(1 + b)$ and $(1 + c)$ must also be integers.
   For $(1 + a)(1 + b)(1 + c) = -5$, the product of the three integers must be -5.
   This is only possible when -5 can be factorized as
   $1 \times 1 \times -5$ or $1 \times -1 \times 5$.
   But, since a, b and c are all different, the only combination is:
   $\qquad 1 + a = 1, 1 + b = -1$ and $1 + c = 5$
   Thus $a = 0$, $b = -2$ and $c = 4$.
   Hence $a + b + c = 0 - 2 + 4 = 2$.

9. $q = 8$.

   Hint: Show that $p = \dfrac{q + 1000}{q - 1} = 1 + \dfrac{1001}{q - 1}$.

   Observe that $1001 = 7 \times 11 \times 13$.

10. 16.

    $x + y + z = 43$
    $10x + 5y + 2z = 229$

    Hints:
    (a) Show that $8x + 3y = 143$
    (b) Observe that $0 \le x < 18$.
    (c) Show that x must be 16.

Mathematical Quickies & Trickies 14 (p. 91)

1. 1998.

   $1999 \times 19{,}981{,}998 - 1998 \times 19{,}991{,}998$
   $= 1999 \times 1998 \times 10{,}001 - 1998(19{,}991{,}999 - 1)$
   $= (1998 \times 1999 \times 10{,}001) - 1998[1999(10{,}001) - 1)]$
   $= (1998 \times 1999 \times 10{,}001) - (1998 \times 1999 \times 10{,}001) + 1998$
   $= 1998$

2. 1.

   The number are the digits of the recurring fraction $\dfrac{1}{7}$.

3. $-2{,}001{,}000$.

   Hint: $a^2 - b^2 = (a - b)(a + b)$.

4. 3535.

   If x represents the number of coins Mr. Ian has, and y represents the number of coins in the second bag, then
   $$\frac{4}{5}x = \frac{y}{7} \times x + 303$$
   $$\frac{x(28 - 5y)}{35} = 3 \times 101$$
   $$x(28 - 5y) = 3 \times 5 \times 7 \times 101$$
   Since $x > 303$ and x divides both 5 and 7, take $x = 5 \times 7 \times 101 = 3535$, then $28 - 5y = 3$, or $y = 5$
   Check: $707 + 2525 + 303 = 3535$.

5. 15 hours.

6. 165.

   $360 = 2 \times 2 \times 2 \times 3 \times 3 \times 5$
   $x = 2 \times 2 \times 2 \times 3 \times 5 = 120$
   $y = 3 \times 3 \times 5 = 45$
   $x - y = 75$; $x + y = 165$.

7. 285.

   Let $a = 15m$ and $b = 15n$
   $a - b = 15(m - n) = 165 = 15 \times 11$
   $m - n = 11$
   $a = 3 \times 5 \times \boxed{3 \times 5} = 225$
   $b = 3 \times 5 \times \boxed{2 \times 2} = 60$
   $a + b = 225 + 60 = 285$.

8. 7.

   Consider the last digits of the powers of 7:
   $7^0 \ 7^1 \ 7^2 \ 7^3 \ 7^4 \ 7^5 \ 7^6 \ 7^7$
   $1 \ \ 7 \ \ 9 \ \ 3 \ \ 1 \ \ 7 \ \ 9 \ \ 3$
   $2009 = 4 \times 502 + 1$
   With a repeating pattern of 4,
   7^{2009} has the same remainder as 7^1, which is 7 when divided by 10.
   Hence, when 7^{2009} is divided by 10, its remainder is 7.

9. $\frac{1}{3}$.

   Let $S = \frac{1}{4} + \frac{1}{16} + \frac{1}{64} + \frac{1}{256} + \ldots$
   Then
   $$S = \left(\frac{1}{2} + \frac{1}{4} + \frac{1}{8} + \frac{1}{16} + \ldots\right) - \left(\frac{1}{2} + \frac{1}{4} + \frac{1}{32} + \frac{1}{128} + \ldots\right)$$
   $$S = 1 - \frac{1}{2}\left(1 + \frac{1}{4} + \frac{1}{16} + \frac{1}{64} + \ldots\right)$$
   $$S = 1 - \frac{1}{2} - \frac{1}{2}\left(\frac{1}{4} + \frac{1}{16} + \frac{1}{64} + \ldots\right)$$
   $$S = \frac{1}{2} - \frac{1}{2}S$$
   $$\frac{3}{2}S = \frac{1}{2}$$
   $$S = \frac{1}{2} - \frac{2}{3} = \frac{1}{3}$$
   Hence $\frac{1}{4} + \frac{1}{16} + \frac{1}{64} + \frac{1}{256} + \ldots = \frac{1}{3}$

   Alternatively,
   $$\frac{1}{2} + \frac{1}{4} + \frac{1}{8} + \frac{1}{16} + \ldots = 1 \rightarrow$$

   $$\frac{1}{4} + \frac{1}{16} + \frac{1}{64} + \frac{1}{256} + \ldots \rightarrow$$

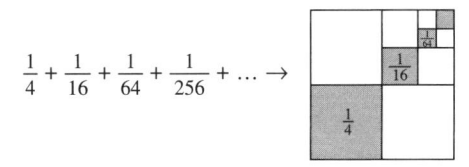

10. 36%.

    48% of the students are not movie addicts and 75% of them are not book lovers.
    Thus, $48\% \times 75\% = 36\%$ are not movie addicts and arre not book lovers.
    Hence the required probability is 36%.

Mathematical Quickies & Trickies 15 (p. 97)

1. 8.

   Spell out each digit, and the last letter of each word is the first letter of the next.
   FIVE → EIGHT → THREE → EIGHT → TWO
   ↓
   ONE

2. 64.

   Take the nth number, add 1 and multiply by n to get the next number each time.
   $n = 1, \quad 0 + 1 = 1, \quad 1 \times 1 = 1$
   $n = 2, \quad 1 + 1 = 2, \quad 2 \times 2 = 4$
   $n = 3, \quad 4 + 1 = 5, \quad 5 \times 3 = 15$
   $n = 4, \quad 15 + 1 = 16, \quad 16 \times 4 = 64$
   $n = 5, \quad 64 + 1 = 65, \quad 65 \times 5 = 325$
   Recursively speaking,
   $f(n + 1) = n[f(n) + 1], f(1) = 0$

3. 103.

   The 12 numbers in the series are linked to the months in the year.

   Write down each month, replace each letter with the number representing its position in the alphabet (so A = 1, B = 2, …) and add up the numbers for each month, and the result will be the numbers shown.

   SEPTEMBER = 19 + 5 + 16 + 20 + 5 + 13 + 2 + 5 + 18 = 103

4. L.

   These are the numbers 0, 1, 2, … typed on a calculator and turned upside-down.

5. 63.

   Take each number, subtract the first digit, and add the second digit, to get the next number.

   75 − 7 = 68, 68 + 5 = 73
   73 − 7 = 66, 66 + 3 = 69
   So, 70 − 7 = 63, 63 + 0 = 63

6. 73, 3000.

   Each of the numbers is the smallest positive whole number for which the name uses a certain number of letters. The next two numbers are 73 and 3000, because SEVENTY-THREE and THREE THOUSAND are the smallest numbers needing 12 and 13 letters respectively. *What comes after 3000?*

7. This is the fraction $\frac{3}{7}$ expressed in decimal form.

8. They represent the last digits of the squares of the counting numbers beginning from 1.

9. 6.

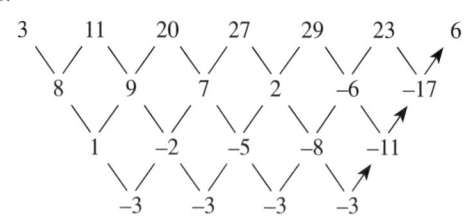

10. $\frac{1}{30}$.

    $12 = \frac{1}{7} \times 84$

    $2 = \frac{1}{6} \times 12$

    $\frac{2}{5} = \frac{1}{5} \times 2$

    $\frac{1}{10} = \frac{1}{4} \times \frac{2}{5}$

    $\frac{1}{30} = \frac{1}{3} \times \frac{1}{10}$

Mathematical Quickies & Trickies 16 (p. 104)

1. 3 mangoes.

   You took 3, so that is how many you have.

2. (a) (1 + 2) ÷ 3 + 4 + 5 = 10
   (b) (1 + 2 + 3 + 4) × 5 = 50
   Are there other solutions?

3. $\underbrace{999\ldots99{,}996}_{24\ 9\text{'s}}$ – A number made up of 24 consecutive 9's followed by a 6.

4. Zero.

   If five are in the right envelope, so is the sixth.

5. 16.

   $x^2 + 3xy + 2y^2 = (x + y)^2 + (x + y)y$
   $= 25 + 5y$
   $y = \frac{40 - 25}{5} = 3$
   $x = 5 - y$
   $x = 2$
   $2x + 4y = 2(2) + 4(3)$
   $= 16$

6. 5 days.

   Each dog digs a hole in 5 days.

7. A solution to the above problem is:

   $\begin{array}{r} 2463 \\ +\ 6518 \\ \hline 8981 \end{array}$

   Is this a unique solution?

   Note: H + N gives a single digit and since a number cannot begin with a 0, so neither H nor N represents 0.

   Next, H and N give an answer E and so do N and S, which means that there must be a carry over of one either from the first to the second column or from the third to the fourth. In other words, there is a difference of one between H and S.

194 More Mathematical Quickies & Trickies – Answers/Hints/Solutions

From the first E of the word EYES we can say that E does not represent 0, 1 or 2. (Why?) With this fact, and noting that N and S appear twice and E three times, we may try figures in various positions.

Proving that any solution is unique is not always an easy matter, since the obvious method of trying to find a second solution, or even third or fourth, may be limited by one's ability rather than by the problem!

9. 4.

    Note that
    $$\frac{1}{\sqrt{n}+\sqrt{n+1}} = \sqrt{n+1}-\sqrt{n}$$
    $$\frac{1}{\sqrt{1}+\sqrt{2}} + \frac{1}{\sqrt{2}+\sqrt{3}} + \ldots + \frac{1}{\sqrt{24}+\sqrt{25}}$$
    $$= \sqrt{2}-\sqrt{1}+\sqrt{3}-\sqrt{2}+\sqrt{4}-\sqrt{3}+\ldots+\sqrt{25}-\sqrt{24}$$
    $$= \sqrt{25}-\sqrt{1}$$
    $$= 4$$

10. Combined, the four dotted hexagons have the same area as the shaded hexagon: If you double the length of the side of a hexagon, you quadruple the area.

Mathematical Quickies & Trickies 17 (p. 113)

1. We need to cut off a piece which is
   $$\frac{2}{3} - \frac{1}{2} = \frac{1}{6} \text{ meter long.}$$
   To find where to make the cut, fold the original piece in half twice; the first yields a piece one-third meter long, and the second yields a piece one-sixth meter long.

   Investigation: What other lengths may be cut off without a ruler?

2. 3.

    Ten, Nine, Eight, Seven, Six, Five, Four, Three, Two, One

3. 50,123 − 49,876 = 247

4. The thickness is $0.01 \times \underbrace{2 \times 2 \times \ldots \times 2}_{50 \text{ times}}$

    ≈ 1.126 × 10 000 000 000 000 cm, which is about 100 000 000 km — this is about two-thirds the Earth Sun distance.

5. 11, 47 and 71.

6. $1\frac{1}{2}$ days

7. 2,100,010,006

8. $\sqrt{.2^{-2}}$ or $(\sqrt{2})^{-2} = \frac{1}{(\sqrt{2})^2} = \frac{1}{.2} = 5$

9. Old is 30 and Young is 18.

10. 96,420 × 87,531 = 8,439,739,020

Mathematical Quickies & Trickies 18 (p. 119)

1. $\sqrt{4-x} = \sqrt{x-6}$

   For real x, $4 - x \geq 0$ and $x - 6 \geq 0$.

   Since the inequalities $x \leq 4$ and $x \geq 6$ cannot be satisfied at the same time, the domain of the equation $\sqrt{4-x} = \sqrt{x-6}$ is the empty set.

   Hence the equation has no solution in $\mathbb{R}$.

2. (a) $(2x+y)(x-2y) = 7 \times 1$
   $$\left.\begin{array}{l} 2x+y=7 \\ x-2y=1 \end{array}\right\} \Rightarrow \begin{array}{l} x=3 \\ y=1 \end{array}$$

   (b) $(x-y)(3x-2y) = 13$
   $$= 1 \times 13 \text{ or } -13 \times -1$$
   $$\left.\begin{array}{l} x-y=1 \\ 3x-2y=13 \end{array}\right\} \Rightarrow \begin{array}{l} x=11 \\ y=10 \end{array}$$
   $$\left.\begin{array}{l} x-y=-13 \\ 3x-2y=-1 \end{array}\right\} \Rightarrow \begin{array}{l} x=25 \\ y=38 \end{array}$$

3. 25,252,525 = 7 × 3,607,503 + 4

   4 days after Sunday is Thursday.
   So Jesus Christ would return on a Thursday.

4. Note that: $2^{125} = 2^5 \times 2^{120}$
   $$= 32 \times (2^{10})^{12}$$
   $$> 32 \times (10^3)^{12}$$
   $$= 32 \times 10^{36}$$

   Thus 2^{125} is larger than 32×10^{36}.

5. 15 shirts.

   He must have at least 15 shirts. Each Friday morning he puts one on, drops off seven, and picks up seven

6. 7.

Only 7 numbers, between 1 and 60, have a number of letters equal to the sum of their digits.

Number	Written	Number of letters	Sum of the digits
4	Four	4	4
16	Sixteen	7	1 + 6 = 7
36	Thirty-six	9	3 + 6 = 9
38	Thirty-eight	11	3 + 8 = 11
45	Forty-five	9	4 + 5 = 9
50	Fifty	5	5 + 0 = 5
54	Fifty-four	9	5 + 4 = 9

7. 12 different paths.

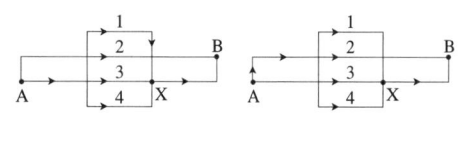

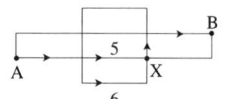

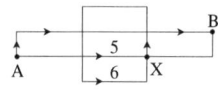

8. Let Joe's parents' ages be x and y when he was born, and let his age when I attended his birthday be d.

   We need to solve: $(x - d)(y - d) = d(2d - 1)$

   If we try $x - d = d$ and $y - d = 2d - 1$ for various values of d we will find the only possible solution in the last hundred or so years.

   Since $976 = 2 \times 26 \times 38$
   and $1989 = 39 \times 51$,
   I must have attended Joe's birthday in 1989 when he was 13.

9. 1.

   Let $p = 1 + \dfrac{1}{2011} = \dfrac{2012}{2011}$

   Then $p \times 201 = 2012$

   $2011 \times p \times p^{2011} = 2012 \times p^{2011}$
   $2011 \times p^{2012} = 2012 \times p^{2011}$

   $a = p^{2011}$ and $b = p^{2012}$

   Hence,
   $\dfrac{a^b}{b^a} = \dfrac{p^{2011 \times p^{2012}}}{2011}$
   $= 1$

10. $x = \dfrac{9}{5}, y = \dfrac{16}{5}$ and $x = -\dfrac{9}{5}, y = -\dfrac{16}{5}$

    $x(x + y) = 9$ (1)
    $y(x + y) = 16$ (2)

    (1) + (2): $x^2 + y^2 + 2xy = 25$
    $(x + y)^2 = 25$
    $x + y = \pm 5$

    If $x + y = -5$, then $-5x = 9 \Rightarrow x = -\dfrac{9}{5}$
    If $x + y = -5$, then $-5y = 16 \Rightarrow y = -\dfrac{16}{5}$
    If $x + y = 5$, then $5x = 9 \Rightarrow x = \dfrac{9}{5}$
    If $x + y = 5$, then $5y = 16 \Rightarrow y = \dfrac{16}{5}$

    Hence the system of equations has two solutions:
    $x = \dfrac{9}{5}, y = \dfrac{16}{5}$ and $x = -\dfrac{9}{5}, y = -\dfrac{16}{5}$

Mathematical Quickies & Trickies 19 (p. 126)

1. 4.
   $\sqrt{x + 5} - \sqrt{x} = 1$
   $\sqrt{x + 5} = 1 + \sqrt{x}$
   $x + 5 = 1 + 2\sqrt{x} + x$
   $\sqrt{x} = 2$
   $x = 4$

2. Remove the 4th and 11th paper clips from the chain.

3. No real x exists.
   $\sqrt{x + 2} = -2$
   The left-hand side is nonnegative and the right has the negative number -2.
   Hence the equation $\sqrt{x + 2} = -2$ has no real root.

4. $0! + 0! + 0! = 1 + 1 + 1 = 3$

5. $b = 8$.
   $ab - a - b = 1000$
   $ab - a - b + 1 = 1001$
   $(a - 1)(b - 1) = 7 \times 11 \times 13$
   Since a is a perfect square, let $a = k^2$ for some integer k.
   Now, $(k^2 - 1)(b - 1) = 7 \times 11 \times 13$
   $(k - 1)(k + 1)(b - 1) = 7 \times 11 \times 13$
   So k must be 12 and $b = 7 + 1 = 8$

196   More Mathematical Quickies & Trickies – Answers/Hints/Solutions

6. $\sqrt{2x+3} + \sqrt{x+3} = 0$

   For $x \geq -\frac{3}{2}$, we have $\sqrt{2x+3} \geq 0$ and $\sqrt{x+3} > 0$;

   for $x < -\frac{3}{2}$, the equation makes no sense.

   The left-hand side of the equation $\sqrt{2x+3} + \sqrt{x+3} = 0$ is a sum of a nonnegative and a positive number, and the left has a value 0.

   Hence the equation has no solution.

7. 2520.

   1 is the smallest integer divisible by 1;
   2 is the smallest integer divisible by 1 and 2;
   6 is the smallest integer divisible by 1, 2 and 4;
   and so on.
   The next number in the sequence, following 840, is 2520, the smallest integer divisible by 1, 2, 3, ..., 9.

8. $\frac{1}{2}, -\frac{3}{2}$.

   Let $t = \sqrt{\frac{3-x}{2+x}} \geq 0$

   Then the equation becomes $t + \frac{3}{t} = 4$

   $t^2 - 4t + 3 = 0$
   $(t-1)(t-3) = 0$
   $t = 1$ or $t = 3$

   If $t = \sqrt{\frac{3-x}{2+x}} = 1$
   $x = \frac{1}{2}$

   If $t = \sqrt{\frac{3-x}{2+x}} = 3$
   $x = -\frac{3}{2}$

   Substituting $x = \frac{1}{2}$ and $x = -\frac{3}{2}$ into the original equation, we have

   $L\left(\frac{1}{2}\right) = 4 = R\left(\frac{1}{2}\right)$ and $L\left(-\frac{3}{2}\right) = 4 = R\left(-\frac{3}{2}\right)$

   Hence the roots of the original equation are $x = \frac{1}{2}$ and $x = -\frac{3}{2}$.

9. Only one politician is honest and 99 are dishonest.

10. 10
    $x^3 - 101x^2 - 999x + 100{,}900 = 0$
    $(x^2 - 101)(x^3 - 999) + 1 = 0$

    Clearly, the only integral solution is $x = 10$.

Mathematical Quickies & Trickies 20 (p. 132)

1. 17,280.

   There are $3! = 3 \times 2 \times 1$ different ways to shop in Singapore, $4! = 4 \times 3 \times 2 \times 1$ ways to shop in Hong Kong, and $5! = 5 \times 4 \times 3 \times 2 \times 1$ ways to shop in Dubai. Hence, there are $3! \times 4! \times 5! = 17{,}280$ different itineraries.

2. 260,000.

   There are 10^3 choices for the first three digits, 26 for the letter, and 10 for the last digit. Hence, there are $10^3 \times 26 \times 10 = 260{,}000$ different licence plates.

3. If Rick were type A, he would not ask the question. If he is type B, the answer is "no," so Ruth must be type A.

4. 2^{100}.

   Each element in the 100-element set can either be included or not included in a subset. Hence, there are $\underbrace{2 \times 2 \times 2 \times \ldots \times 2}_{100} = 2^{100}$

5. 333

   $2^n > 10^{100}$
   $n \log 2 > 100 \log 100$
   $n \log 2 > 100$
   $n > \frac{100}{\log 2} \approx 333$

6. Yes.

   Instead of one monk climbing the hill one day and returning the next, imagine two men, one going up and one coming down on the same day, both on the same path and walking at the same speed as the monk. Where they meet is the place that the monk passed at the same time going both ways.

7. There are $3^{10} = 59{,}049$ such strings.

8. There are $4! (= 4 \times 3 \times 2 \times 1)$ ways to order the science books, $5! (= 5 \times 4 \times 3 \times 2 \times 1)$ ways to order the fiction books, and $6! (= 6 \times 5 \times 4 \times 3 \times 2 \times 1)$ ways to order the history books.

   The groups of books can be ordered in $3!$ ways. Hence, there are $4!5!6!3! = 12{,}441{,}600$ orderings of all the books.

9. There are three possible solutions.

$$\begin{array}{r}9604\\+\ 9604\\\hline 19\ 208\end{array}\qquad\begin{array}{r}9703\\+\ 9703\\\hline 19\ 406\end{array}\qquad\begin{array}{r}9802\\+\ 9802\\\hline 19\ 604\end{array}$$

Note that H + H → H and L + L → L, only 0 + 0 = 0 and 9 + 9 + 1 (carry) = 19 will do.

Since the numbers cannot start with 0, H must be 9 and L must be 0.

10. The fisherman's speed relative to the water is constant. Since he rowed away from the log upstream for an hour, he rowed downstream for an hour before he returned to it. During those two hours the log moved 2 km, so its speed relative to the land was 1 km/h.

Mathematical Quickies & Trickies 21 (p. 138)

1.

2	5	10	17	26	37
50	65	82	101	122	145

All the numbers are one more than the squares.

2.

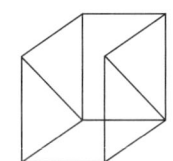

3. If Dave and Esther are 10 km apart after t hours, then there are two cases to consider: (1) Before they meet, (2) After they meet.

Case 1: Before they meet

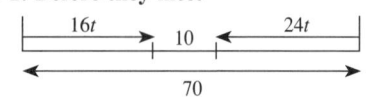

$$16t + 10 + 24t = 70$$
$$40t = 60$$
$$t = 1.5$$

They are 10 km apart the first time after 1.5 hours.

Case 2: After they meet

$$16t + 24t - 10 = 70$$
$$40t = 80$$
$$t = 2$$

They are 10 km apart the second time after 2 hours.

4. Let $n = 10a + b$ with $a, b\ \varepsilon\{0, 1, 2, \ldots, 9\}$ and $a \neq 0$.

Thus $10a + b = 2ab \Rightarrow b = \dfrac{10a}{2a - 1}$

Hence $(a, b) = (3, 6)$ is obtained, i.e. $n = 36$ is true.

5. Note that one of the toys comes from the first box we buy.

The likelihood or chance of getting a new toy from the next box is $\dfrac{3}{4}$, on average; therefore, we would need to buy $1 \div \dfrac{3}{4} = \dfrac{4}{3}$ boxes to get a new toy.

The third new toy would require an additional $1 \div \dfrac{1}{2} = 2$ boxes.

The fourth would require, on average, an additional $1 \div \dfrac{1}{4} = 4$ boxes.

In total, the average number of boxes is

$1 + \dfrac{4}{3} + 2 + 4 = 8\dfrac{1}{3}$ boxes

Note: In practice, we cannot buy $\dfrac{1}{3}$ of a box, but that is still the average number of boxes you would have to buy.

6. The area in each case is 192 cm².

The maximum area results when two ends form a right angle, at which point the area is 200 cm². Note that at this maximum point, some kind of symmetry is at play.

7. 12:00 and 11:20.

8. Let the initial number of stations be f, and the number of new stations be n.

Then $\quad 2fn + n(n - 1) = 52$
$\qquad n^2 + (2f - 1)n - 52 = 0$

Product of roots $= -52 = n_1 \times n_2$
Sum of roots $= n_1 + n_2 = -(2f - 1)$
Since $52 = 1 \times 52$
$\qquad = 2 \times 26$
$\qquad = 4 \times 13$,

the sum of the roots must equal 51, –51, 24, –24, 9, or –9.

But this sum must equal $2f - 1$, with f a positive integer.

So the only possibility is $2f - 1 = 9$, which gives $f = 5$.

The equation then factors as $n^2 + 9n - 52 = 0$,
$\qquad\qquad\qquad\qquad\qquad (n - 4)(n + 13) = 0$

Since $n > 0$, $n = 4$

So there were 5 stations initially, and 4 were added.

Note: If $2f - 1 = 51$, then $f = 26$ and $n = 1$. But $n > 1$.

9. (192, 384, 576), (219, 438, 657), (273, 546, 819), (327, 654, 981)

10. 10201.
    $10201 = 101^2$
    $\sqrt{104060401} = 10201$

Mathematical Quickies & Trickies 22 (p. 146)

1. 7.5 s — it tales 1.5 s between each strike.

2. By the product rule, there are $3 \times 4 \times 2 = 24$ choices.

3. 15 silver coins.
   5 weights can balance either 25 silver coins or 30 gold coins.
   Since 12 gold coins are used, this means the weight of $(30 - 12) = 18$ gold coins is to be used by the silver coins.
   6 gold coins → 5 silver coins
   18 gold coins → 15 silver coins

4. (a) Friday (b) Sunday
   In normal days, each date is one day further on in the week than it was the year before. This is because 365 days make 52 weeks and 1 day.
   The second year, 1904, was a leap year (366 days). This extra day in 1904 makes the gap two days further on into the week, compared with the previous year.

5. The number of arrangement is $\frac{(5+4+3)!}{5!4!3!}$
   $= 27,720$.

6. 70.

7. $5\sqrt{31,770}$, $6\sqrt{18,990}$, $7\sqrt{19,002}$, $8\sqrt{35,112}$.

8. From the collection of 2^{26} subsets of $\{a, b, c, \ldots, x, y, z\}$, we must remove those in which the letters c, h, r, i, s, t appear. Since there are 2^{20} of these subsets, the number of desired subsets is $2^{26} - 2^{20} = 66,060,288$.

9. 60 cents.
   $\frac{96-x}{x} \times 100 = x$
   $x^2 + 100x - 9600 = 0$
   $(x + 160)(x - 60) = 0$
   Since $x > 0$, $x = 60$.

10. 2.
    $1 + \frac{1}{3} + \frac{1}{6} + \frac{1}{10} + \frac{1}{15} + \cdots$
    $= 2\left(\frac{1}{1 \times 2}\right) + 2\left(\frac{1}{2 \times 3}\right) + 2\left(\frac{1}{3 \times 4}\right) + 2\left(\frac{1}{4 \times 5}\right) + \cdots$
    $= 2\left(1 - \frac{1}{2}\right) + 2\left(\frac{1}{2} - \frac{1}{3}\right) + 2\left(\frac{1}{3} - \frac{1}{4}\right) + 2\left(\frac{1}{4} - \frac{1}{5}\right) + \cdots$
    $= 2\left(1 - \frac{1}{2} + \frac{1}{2} - \frac{1}{3} + \frac{1}{3} - \frac{1}{4} + \frac{1}{4} - \frac{1}{5} + \cdots\right)$
    $= 2 \times 1$
    $= 2$

Mathematical Quickies & Trickies 23 (p. 152)

1. 300 km.

2. You cannot dig half a hole!

3. 23 and 24.

4. $55^2 = 3025$; $99^2 = 9801$.

5. $n = b - a - 1$.
   The pages numbered (1, 2), (3, 4), (5, 6), ..., (right, left), ... The last numbered page *before* the torn-out section is a left-hand page and will have an even number. The first numbered page *after* the torn-out section will be a right-hand page and will have an odd number.
   Finding the difference of a and b counts every missing page number but one. Thus, the difference (an odd number) minus one gives the number of missing page numbers.

6. 501.
   A sequence of numbers that has a remainder of 4 when divided by 7 is as follows:
   4, 11, 18, 25, 32, 39, 46, 53, 60, ...
   The next sequence of numbers giving a remainder of 5 when divided by 8 is:
   5, 13, 21, 29, 37, 45, 53, 61, 69, ...
   The least integer meeting the first two conditions is 53. Continuing both sequences shows that the next such integer is 109.
   Notice that 53 is 3 less than the LCM of 7 and 8. Since 501 is 3 less than the LCM of 7, 8, and 9, is it a desired solution?
   The following list shows numbers that give a remainder of 6 when divided by 9 is:
   6, 15, 24, 33, 42, 51, 60, 69, 78, ...

More Mathematical Quickies & Trickies – Answers/Hints/Solutions **199**

Comparing the numbers in this list to those in the previous two lists shows that the "LCM – 3" theory holds ($7 \times 9 - 3 = 60$ and $8 \times 9 - 3 = 69$).

Extension: What about adding these extra conditions:
(d) dividing by 6 gives a remainder of 3,
(e) dividing by 11 gives a remainder of 8?

Making complete lists of numbers is tedious. The number is LCM (6, 7, 8, 9, 11) – 3
$= 2^2 \times 3^2 \times 7 \times 11 - 3 = 5541$.

7. $647 \times 9 = 5823$.

8. $(5 - 2) \times [1 + (4 \div 6)] = 5$.

9. There cannot be more than 50 integers in the group or there will be some pair totaling 100. There are a number of solutions of 50 integers, of which the simplest is the group of integers 50 through 99.

10. Since $3^2 + 4^2 = 5^2$ and $5^2 + 12^2 = 13^2$, we have
$$3^2 + 4^2 + 12^2 = 13^2$$

Mathematical Quickies & Trickies 24 (p. 157)

1. Friday.
$40 = 7 \times 3 + 5$
5 days after Sunday is Friday.

2. 10.
Sum of the three numbers below the diameter
$= \frac{1}{4} \times$ top number

3. $55 = 44 + \frac{44}{4}$.

4. 6, 10.
$10^2 - 6^2 = 100 - 36 = 64 = 4^3$
$10^3 - 6^3 = 1000 - 216 = 784 = 28^2$

5. 773.

6. Jeremy is 3 and his father is 51.
$3 \times 51 = 153$

7. 84.
$x = \frac{1}{6}x + \frac{1}{12}x + \frac{1}{7}x + 5 + \frac{1}{2}x + 4$
On solving, $x = 84$.

8. $9510 + \frac{738}{246} = 9513$.

9. $9^{21} = 109{,}418{,}989{,}131{,}512{,}359{,}209$
For values of n larger than 21, 10^n has $n + 1$ digits, while 9^n has less than n digits.

10. The smallest number that satisfies these conditions is 35,641,667,749. Other numbers may be obtained by adding 46,895,573,610, or any multiple of it.

Mathematical Quickies & Trickies 25 (p. 162)

1.
1	2	4	6	10	12	16	18
22	28	30	36	40	42	46	52

2. $\sqrt{2}$.
$(a + b)^2 = a^2 + b^2 + 2ab$
$= 6ab + 2ab$
$= 8ab$
$(a - b)^2 = (a + b)^2 - 4ab$
$= 8ab - 4ab$
$= 4ab$
$\frac{a + b}{a - b} = \frac{\sqrt{8}\sqrt{ab}}{2\sqrt{ab}} = \frac{2\sqrt{2}}{2} = \sqrt{2}$

3. Since $pq = p + q$,
$pq - p - q + 1 = 1$
$(p - 1)(q - 1) = 1$
We can let $p = \alpha \neq 1$ and $q = 1 + \frac{1}{\alpha - 1}$
$q = \frac{\alpha}{\alpha - 1}$ for any $1 \neq a \in \mathbb{R}$

4. 10, 6.
Let $a - b = k$, $a + b = 4k$, and $ab = 15k$
$a - b = k$
$a + b = 4k$
$\overline{\quad 2a = 5k}$
$a = \frac{5k}{2}$
$b = a - k$
$= \frac{5k}{2} - k$
$= \frac{3k}{2}$
Now, $ab = \left(\frac{5k}{2}\right)\left(\frac{3k}{2}\right) = 15$
$15k = \frac{3k}{2} = 15$
$k^2 - 4k = 0$
$k(k - 4) = 0$
$k = 0$ or $k = 4$
Ignoring $k = 0$, we have $k = 4$.

Thus, $a = \frac{5k}{2} = \frac{5}{2} \times 4 = 10$ and
$b = \frac{3k}{2} = \frac{3}{2} \times 4 = 6$

Check: $a + b$: $a + b$: ab
  4 : 16 : 60
  1 : 4 : 15

5. $500 = 444 + 44 + 4 + 4 + 4$.

6. One of the answers is $8463 \div 7 = 1209$.

7. $a = \frac{14}{3}, b = \frac{10}{3}, c = 6$

Let $2a + 2b + 2c = 6k + 7k + 8k$
$\qquad\qquad\qquad\quad = 21k$
$a = \frac{21}{2}k - 7k = 3.5k$
$b = \frac{21}{2}k - 8k = 2.5k$
$c = \frac{21}{2}k - 6k = 4.5k$

Now $a + b + c = 14$
$\Rightarrow \frac{21}{2}k = 14$
$k = 14 \times \frac{2}{21} = \frac{4}{3}$
$a = \frac{7}{2} \times \frac{4}{3} = \frac{14}{3}$
$b = \frac{5}{2} \times \frac{4}{3} = \frac{10}{3}$
$c = \frac{9}{2} \times \frac{4}{3} = 6$

8. $299 = 297 + \frac{1086}{543}$.

9. $21{,}978 \times 4 = 87{,}912$

(i) Since 4 is even, a must be even.
But the product still has 5 digits.
We must have $a = 2$. This implies $e = 3$ or $e = 8$.
Since $4a = e$, we must have $e = 8$, i.e.,
$\qquad 2bcd8$
$\qquad \times \quad 4$
$\qquad \overline{8dcb2}$
is true.

(ii) b must be ≤ 2. Otherwise $4b > 10$ and the first digit of the product cannot be 8.
If $b = 2$, then
$\qquad\quad d8$
$\qquad \times\ 4$
$\qquad \overline{\square 22}$
cannot be satisfied. Similarly, $b = 0$ is not correct either.

Hence $b = 1$ is obtained, i.e.
$\qquad 21bcd8$
$\qquad \times \qquad 4$
$\qquad \overline{8dcb12}$

10. Let $\angle c = 90°$
$a^2 + b^2 = c^2$
$\frac{1}{2}ab = 2(a + b + c)$
$a + b + c = \frac{ab}{4}$
$c = \frac{ab}{4} - (a + b) = \frac{ab}{4}$
$a^2 + b^2 = \frac{ab}{4}$
$\qquad = a^2b^2 - 8a^2b - 8ab^2 + 16a^2 + 32ab + 16b^2$
$ab(ab - 8a - 8b + 32) = 0$
$ab \neq 0 \Rightarrow ab - 8a - 8b + 3 = 0$
$\qquad\qquad ab - 8a - 8b + 64 = 32$
$\qquad\qquad (a - 8)(b - 8) = 32$

Thus, $a - 8 = 1, b - 8 = 32$, or
$\qquad a - 8 = 2, b - 8 = 16$, or
$\qquad c - 8 = 4, b - 8 = 8$ are true

If $a = 9, b = 40, c = \sqrt{9^2 + 40^2} = 41$

If $a = 10, b = 24, c = \sqrt{10^2 + 24^2} = 26$

If $a = 12, b = 16, c = \sqrt{12^2 + 16^2} = 20$

Hence there exist only three such right triangles with side-lengths of (9, 40, 41), (10, 24, 26) and (12, 16, 20).

Mathematical Quickies & Trickies 26 (p. 168)

1. Good with numbers.

2. 25.

Note that $\left(\frac{a+b}{2}\right)^2 - ab = \left(\frac{a-b}{2}\right)^2 = 25$
$\rightarrow (a - b)^2 = 25 \times 4$
$\qquad a - b = \pm 10$
$\qquad\qquad b = a - 10$ or
$\qquad\qquad b = a + 10$

$\frac{a^2 + b^2}{2} - \left(\frac{a+b}{2}\right)^2$
$= \frac{2a^2 \mp 20a + 100}{2} - (a^2 \mp 10a + 25)$
$= 25$

3. $1, -3 \pm 2\sqrt{2}$.

   Let $y = x + \dfrac{1}{x}$.

   Then $\left(x^2 + \dfrac{1}{x^2}\right) + 4\left(x + \dfrac{1}{x}\right) - 10$

   $= \left(x + \dfrac{1}{x}\right)^2 - 2 + 4y - 10$

   $= y^2 + 4y - 12 = 0$, which implies

   $y = 2$ or $y = -6$

   If $y = 2$, then $x^2 - 2x + 1 = 0$ implies $x = 1$

   If $y = -6$, then $x^2 + 6x + 1 = 0$ implies $x = -3 \pm 2\sqrt{2}$.

4. Let $n = 10a + b$ with $a, b \in \{1, 2, ..., 9\}$,

   then $10b + a = \dfrac{7}{4}(10a + b) \Rightarrow b = 2a$.

   Thus $n = 12$, or 24, or 36, or 48 are obtained.

5. 41.

   Let the original integer be ab,
   $ab = 10a + b$ with $a, b \in \{0, 1, 2, ..., 9\}$ and $a \neq 0$.

   Thus
   ```
     a b 3         3 7 2
   - 3 7 2       + a 1
   ─────────     ─────────
     a b ⇒ b = 1,  a 1 3 ⇒ a = 4,
   ```
   i.e., the given integer is 41.

6. 12.

   The last digit of n must be 2.
   If the last second digit of n is 0, then the last second digit of the product must be 9.
   Hence the last second digit of n must be 1.
   But $999 \times 12 = 11{,}988$ is not the required answer.
   We try $n = 112$.
   Hence $999 \times 112 = 111{,}888$ is obtained.

7. If $x = 1$, then $y = x^2 + 1 = 2$

   $\Rightarrow 4 = y^2 \neq 1 - y$

   i.e. $x \neq 1$ is true.

   Thus $y = \dfrac{x^2 + 1}{x}$

   $= x + \dfrac{1}{x}$

   $\Rightarrow x^2 + 2 + \dfrac{1}{x^2} = y^2$

   $= 1 - y$

   $= 1 - x - \dfrac{1}{x}$

   and $x^4 + x^3 + x^2 + x + 1 = 0$

   This shows $x^5 = 1$.

8. Note that $\left(x - \dfrac{1}{x}\right)^2 = \left(x + \dfrac{1}{x}\right)^2 - 4 = 0$ and $x = \pm 1$.

   This implies $x^{2n} + \dfrac{1}{x^{2n}} = 2$ and $x^{2n-1} - \dfrac{1}{x^{2n-1}} = 0$ for any positive integer n.

9. Each column is the addition A + B + C, but the sums are different. Hence we have

   $10 \leq A + B + C \leq 7 + 8 + 9 = 24$.

   If $A + B + C \geq 20$, then $J - K = 2$ and $I = J$ are true. Hence we further conclude $10 \leq A + B + C \leq 19$. We can replace $A + B + C = 19$ is correct.
   Hence $(A, B, C) = (4, 7, 8), (5, 6, 8)$ are true, i.e.,

   ```
     444         555
     777         666
   + 888       + 777
   ─────       ─────
    2109        2109
   ```

10. Clearly, n has 3 digits.
    The first digit is 1 and the last digit is 3 or 7, i.e. $n = 1 \square 3$ or $n = 1 \square 7$.
    By putting digits 0 ~ 9 in the box and using a calculator, then $n = 167^2$, $x = 7$, $y = 8$ are obtained.

Mathematical Quickies & Trickies 27 (p. 175)

1. (a)

1	2	4	6	10	12	16	18
22	28	30	36	40	42	46	52

   All the numbers are one less the primes.

   (c)

2	5	10	17	26	37
50	65	82	101	122	145

   All the numbers are one more the the squares.

2. $500 + 444 + 44 + 4 + 4 + 4$.

3. $299 = 297 + \dfrac{1086}{543}$.

4. $2451 = \underbrace{3 \times 19}_{57} \times 43$ and $57 + 43 = 100$

   Hence, the two numbers are 43 and 57.

5. Clearly, $\dfrac{1}{n} = \dfrac{1}{n+a} + \dfrac{1}{n+b}$ iff $ab = n^2$ is true.

   Thus we can let $p = n + a$ and $q = n + b$.

6. 4 to 3.

   There are 12 equilateral triangles in the 6-pointed star.
   There are 9 equilateral triangles in one large triangle.
   Thus, the ratio is $\dfrac{12}{9} = \dfrac{4}{3}$.

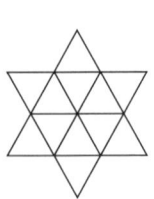

Alternatively,
Star = 1 large equilateral + 3 small equilateral triangles with sides one-third the size of the large triangle's sides.

Area of each small triangle
= $\frac{1}{9}$ × area of the large triangle.
The ratio is $1 + \frac{1}{9} \times 3 = \frac{4}{3}$.

7. 0.
$$x^3 + \frac{1}{x^3} = \left(x + \frac{1}{x}\right)^3 - 3x - \frac{1}{x}\left(x + \frac{1}{x}\right)$$
$$= \left(x + \frac{1}{x}\right)\left[\left(x + \frac{1}{x}\right)^2 - 3\right]$$
$$= 0$$

8. $\frac{1}{x} + \frac{1}{x} + \frac{1}{x} = \frac{1}{x} = (x+y)(y+z)(x+z) = 0$
Hence $\begin{cases} x = -y \\ z = w, \end{cases}$ $\begin{cases} y = -z \\ x = w, \end{cases}$ $\begin{cases} z = -x \\ y = w \end{cases}$
are the solutions of the given system.

9. Let a and b be the length and width of the rectangle.

Then $ab = 6(a+b)$
$ab - 6a - 6b + 36 = 36$
$(a-6)(b-6) = 36$

Let's consider the following cases:
(i) $a - 6 = 1, b - 6 = 36 \Rightarrow a = 7, b = 42$
(ii) $a - 6 = 2, b - 6 = 18 \Rightarrow a = 8, b = 24$
(iii) $a - 6 = 3, b - 6 = 12 \Rightarrow a = 9, b = 18$
(iv) $a - 6 = 4, b - 6 = 9 \Rightarrow a = 10, b = 15$
(v) $a - 6 = 6, b - 6 = 6 \Rightarrow a = 12, b = 12$
(vi) $a - 6 = -1, b - 6 = -36 \Rightarrow a = 5, b = -30$
(vii) $a - 6 = -2, b - 6 = -18 \Rightarrow a = 4, b = -12$
(viii) $a - 6 = -3, b - 6 = -12 \Rightarrow a = 3, b = -6$
(ix) $a - 6 = -4, b - 6 = -9 \Rightarrow a = 2, b = -3$
(x) $a - 6 = -6, b - 6 = -6 \Rightarrow a = 0, b = 0$

Hence the possible areas are
$42 \times 7 = 294$, $24 \times 8 = 192$, $18 \times 9 = 162$,
$15 \times 10 = 150$ and $12 \times 12 = 144$ square units.

10.

Equilateral triangles	Isosceles triangles	Scalene triangles
1-1-1	1-2-2	2-3-4
2-2-2	1-3-3	2-4-5
3-3-3	1-4-4	3-4-5
4-4-4	1-5-5	
	2-3-3	
	3-4-4	
	2-5-5	
	3-4-4	

More Mathematical Quickies & Trickies
Bibliography & References

Andreescu, T., Andrica, D. & Feng, Z. (2007). *104 number theory problems: From the training of the USA IMO team*. Boston: Birkhauser.

Andreescu, T. & Feng, Z. (2005). *103 trigonometric problems: From the training of the USA IMO team*. Boston: Birkhauser.

Beiler, Albert H. (1966). *Recreations in the theory of numbers: The queen of mathematics entertains*. New York: Dover Publications, Inc.

Bellos, A. (2010). Mathemagical. *New Scientist*, 29 May 2010. pp. 44-49.

Booth-Jones, C. (1978). *More brain ticklers*. Beaver Books.

Carter, P. & Russell K. (2009). *The complete book of fun maths*. West Sussex: John Wiles.

Fisher, L. & Medigovich, W. (1981). *Problem of the week*. Palo Alto, CA: Dale Seymour Publications.

Friedland, A. J. (1970). *Puzzles in math and logic*. New York: Dover Publications, Inc.

Gardiner, A. (1987). *Mathematical puzzling*. New York: Dover Publications, Inc.

Gardner, M. (2008). *Hexagons, probability paradoxes, and the Tower of Hanoi*. New York: Cambridge University Press.

Gardner, M. (2006). *The colossal book of short puzzles and problems*. New York: W. W. Norton & Company Ltd.

Goldwich, David (2009). *Getting into Singapore: A guide for expats and kaypoh Singaporeans*. Singapore: David Goldwich.

Harshman, E. J. (2002). *Mind-sharpening lateral thinking puzzles*. New Delhi: Orient Paperbacks.

Hunter, J. A. H. & Madachy, Joseph S. (1963). *Mathematical diversions*. New York: Dover Publications, Inc.

Mott-Smith, Geoffrey (1954). *Mathematical puzzles for beginners and enthusiasts*. New York: Dover Publications, Inc.

Mack, D. R. (1992). *The unofficial IEEE brainbuster game book*. New York: The Institute of Electrical and Electronics Engineers, Inc.

Pickering David (2006). *Penguin Pocket Jokes*. Penguin Books.

Stickels, T. (2003). *Mesmerizing mind-bending puzzles*. New Delhi: Orient Paperbacks.

Stickels, Terry (2000). *Are you as smart as you think?* New York: St. Martin's Press.

Sundem, G. (2009). *The geeks' guide to world domination*. New York: Random House.

Vennebush, G. P. (2010). *Math jokes 4 mathy folks*. Bandon, OR: Robert D. Reed Publishers.

Yan, K. C. (2011). *Mathematical quickies & trickies (expanded & updated ed.)*. Singapore: MathPlus Publishing.

Yan, K. C. (2009). *Geometrical quickies & trickies*. Singapore: GLM Pte Ltd.

Yan, K. C. (2009). *Get calculator smart*. Singapore: EPB Pan Pacific.

Yan, K. C. (2008). *Mind Stretchers 2*. Singapore: Panpac Education.

Yan, K. C. (1998). Mathematical Mantras. *The Mathematics Educator*, 3(1), 101-11.